LOTUS SINCE THE 70s
Volume 2: Esprit, Etna and V8 engines

LOTUS SINCE THE 70s

Volume 2: Esprit, Etna and V8 engines

A collector's guide
by Graham Robson

MOTOR RACING PUBLICATIONS LTD
Unit 6, The Pilton Estate, 46 Pitlake, Croydon CR0 3RY, England

First published 1993

British Library Cataloguing in Publication Data

Robson, Graham
    Lotus Since the 70's: Collector's Guide. –
    Vol. 2: Esprit, Etna and V8 engines
    I. Title
    629.222

ISBN 0-947981-69 1

Part of this book was previously published in 1983 by Motor Racing Publications in the Collector's Guide
The Third Generation Lotuses, which this book and its companion volume (ISBN 0 947981 70 5) supersedes.

Typeset by Ryburn Publishing Services, Keele University, Staffordshire

Printed in Great Britain by
The Amadeus Press Ltd, Huddersfield, West Yorkshire

Contents

Introduction

More than 10 years ago I wrote my first book about Lotus cars (*The Third Generation Lotuses*), which covered all the Hethel-built cars that had been equipped with the famous 16-valve twin-cam engine. Since then, the production lines at Hethel have seen much change.

Therefore, I thought the time was ripe to update the story to the early Nineties, and the point at which the technically-brave front-wheel-drive Elan project was ended. To give adequate coverage to all the different models involved, I was convinced that there was enough subject matter to fill two new books.

I was delighted to learn that my publisher, John Blunsden, shared my enthusiasm for the modern Lotus story, and that he agreed to publish a matched pair of books, one covering all the modern front-engined Lotus cars, and this one the mid-engined cars. Between them, these two books embrace a fascinating 20-year period in which Lotus and their business changed considerably.

At the time of writing, the story of the Esprit ('post-Elan') was still unfolding, and it may be that there are more surprises, more developments and more technical triumphs to come before the end of its long and exciting career. It may even be that versions will appear which do not use the famous 16-valve engine.

Although the original Esprit was conceived in 1972 and unveiled in 1975, the Giugiaro-styled car was on the market for 11 years, and its Stevens-styled successor took over very successfully thereafter. Both were cars that were persistently updated and improved, which continued to sell when by all conventional measures they should have gracefully died away, and which continued to amaze the pundits.

This, then, is 'the story so far'. Even so, I believe it is a comprehensive chronicle of the Esprit's first 20 years of life.

Together with its companion volume, it is certainly the first book to cover in full the often tempestuous period of Lotus production through the Seventies and Eighties and into the Nineties.

My own judgment is that the cars covered here have done much to keep the Lotus marque in business, for during their life Lotus went through several changes of ownership, swung dramatically from profitability to loss, and came very close to financial disaster on more than one occasion.

All the derivatives were brave in concept, sophisticated in engineering, their specification so advanced and high-tech that prices were always bound to be high and sales limited. Yet when Colin Chapman conceived the Esprit early in the Seventies he could not possibly have guessed that two severe energy crises and several major recessions would follow during its long career.

Commercially, therefore, the Esprit was a big gamble when it was being developed, and was certain to be a costly car when put on sale, yet this never deterred Lotus from forging ahead. A cautious business might never have taken this gamble, nor have persevered with it for so long in the face of a very worrying series of financial traumas.

Lotus enthusiasts, however, tend to be far more interested in performance, character and style than in the study of a balance sheet, so I should stress that each and every one of the cars covered in this book was an instant 'classic', and their long-term future looks to be assured.

I hope the information, opinion and pictures included here will help all Lotus owners to enjoy and preserve their cars in the years to come.

February 1993

Graham Robson

Acknowledgements

In assembling material for this book, I have often had to draw on my links with Lotus, which go back many years.

I want to thank Mike Kimberley, Managing Director at Hethel during the Eighties – and later Chairman of the company – not only for finding the time to be interviewed, but for allowing me to disrupt the smooth running of parts of the business, and to dig through the company's archive material for facts and figures.

Over the years I was particularly helped by two keepers of Lotus' public image – Don McLaughlan in the Eighties and Patrick Peal in the Nineties – who both provided statistics, open doors to various personalities, and gave access to the copious pictorial stores. They also put me in touch for data with Sid Harper, Hugh Wilson, Andrew Walmsley and Brian Angus.

Many of the photographs come from Lotus' own archive, and from Focalpoint in Norwich, which is so capably and cheerfully run by Ron and Brenda Middleton. Other photographs have been provided by Aerospace Publishing, Publications International Ltd of the USA, Ray Hutton, my personal North American connection Richard Langworth and by my fellow-author Mark Hughes.

For corroborative facts on the Lotus tie-up with Talbot, my thanks go to Colin Cook, then of the Talbot Motor Co Ltd, for help on American comment regarding Lotus, to the Reference Library of the National Motor Museum at Beaulieu, and to Jeremy Walton for his original research into Esprit matters.

In addition, it would simply not have been possible to put all this into perspective without consulting Miles Wilkins of Fibreglass Services, who probably knows more about classic Lotus cars and their restoration than anyone else in the world. He knows, and you may never realize it, just how much he contributed to this boom.

As ever, I am grateful to *Autocar & Motor*, its former editor Shaun Campbell, and to his erstwhile colleagues over the years when these two magazines were still separate, and to *Thoroughbred & Classic Cars*, its editor Tony Dron and his associate Lionel Burrell, not only for producing the most reliable magazines of record, but for allowing me to quote from their back numbers.

Finally, of course, I salute Lotus-watchers all round the world, not only for their interest in the marque, but for the way they always encouraged Lotus to be more advanced, more inventive and technically more successful with every new model that appeared.

Graham Robson

The Lotus Esprit has travelled a long way since the S1 entered production in 1976, but this 1993 S4 remains loyal to the original concept laid down in the early Seventies. The Esprit remains a supercar in both spirit and performance.

CHAPTER 1

Ancestors and relatives

Elan, Plus 2 and Europa

The story of Lotus is well known – but still remarkable. In essence, it is the story of one man's genius and enterprise, and it covers the evolution of a famous marque through trials specials, kit-cars, racing sportscars, Grand Prix machinery and a series of exciting road cars. The first Lotus was built in 1947, the first sold to a customer in 1953, and the company has grown more famous with every succeeding year.

There is no space in this book to relate the many Lotus racing successes, except to point out that without the racing the company might not have become as well known, and that without the technical stimulus of racing the road cars might not have become as advanced. It ought to be made clear, incidentally, that the connection between Lotus Cars and Team Lotus has been very tenuous indeed for many years – a fact of financial and commercial life which the Group Lotus annual report always spells out with care.

In broad terms, there have been four different families of Lotus road car, though in the context of this book I propose to dismiss the amazing Lotus Seven as a racing sportscar which could be used on the road – Dennis Ortenburger, Lotus Seven authority, will understand I hope! The first civilized Lotus road car, therefore, was the original Elite coupe of 1957–63, while the next complete family was the Elan/Plus 2/Europa series of 1962–75.

Colin Chapman had started supplying Lotus 6 kit-cars in 1953 from a stable in Tottenham Lane, Hornsey, North London. His company had been founded with the aid of a £25 loan from his girlfriend Hazel – who subsequently became his wife – and at this time he was still a salaried employee of the British Aluminium Company. He had already started to race his own products with some success, and his rise to fame was so rapid that for 1956 he was commissioned to design a new chassis space-frame for the Vanwall Grand Prix car. In 1957 he introduced his first Lotus single-seater – the Type 12 Coventry Climax-engined Formula 2 car but in the same year he launched the original Elite.

Strictly speaking, the Type 14 Lotus, which became the Elite, was conceived with an eye to new Grand Touring car race regulations for 1,300cc cars, which explains why it was laid out more with an eye to function than to refinement and comfort. During 1957, it was Colin Chapman who designed the unique glassfibre monocoque type of construction and elected to use a single-cam Coventry Climax engine, while it was his accountant friend Peter Kirwan-Taylor, and Ron Frayling, who set about styling the car.

It is now well known that the Elite made its first public appearance before it had even been completed, let alone driven and developed, that it did not actually get into production until the end of 1958, also that it proved far too expensive to manufacture. Between 1958 and 1963, a total of 1,030 body/chassis units were built at a new Lotus factory in Cheshunt, a few miles further north on the outskirts of London; it is generally agreed that nearly 1,000 of these were built up as complete – or complete kit- – cars.

However, although the Elite was well-loved, usually from a distance by people who could not afford to buy one for themselves, it was not a commercial success. No less an authority than Colin Chapman was once quoted as saying:

The first Lotus road car to use a backbone chassis-frame was the Elan of 1962, and a lengthened version of this design was used for the Elan Plus 2 of 1967. The frame illustrated is that of the Plus 2. The principles of this frame were retained into the Eighties.

'The Elite was really a road-going race car and used many of the racing components. We didn't have much experience of road-going economics when we designed it, and without long-range tooling, long-range buying and strict cost-saving it was finally just uneconomic to build. I believe we lost over £100 on each car we built. Something had to be done, and so we started work on the Elan.'

The Elan was almost everything that the Elite was not: it was a convertible where the Elite had been a coupe; it had a separate chassis and Lotus' own brand of engine; it was refined, habitable and, above all, it was profitable. It helped Lotus forge strong links with Ford, links which they retained into the Seventies, and it represented a big step along the road to respectability as manufacturers rather than as mere assemblers of cars.

The two most important features in the definitive Elan were its separate, pressed-steel backbone chassis-frame and its Lotus-Ford twin-overhead-camshaft engine. The concept of the backbone frame, once adopted for Lotus road cars, has never been abandoned, while the engine programme was merely the first step towards making Lotus independent of engine supplies from any other concern.

In the beginning, it seems, there had been thoughts about building the Elan in unit-construction form, but as it was destined to be an open sportscar in its original form this scheme was speedily dropped. As refined for series production, there was a backbone frame in pressed steel, bifurcated at the front to pass at each side of the front-mounted engine and gearbox assembly, and supporting steel 'towers' to pick up the combined coil spring/damper units at the rear. The glassfibre bodyshell, which had neat and simple styling, influenced once again by Peter Kirwan-Taylor, featured flip-up headlamps in the front wings, had Ford bumpers neatly moulded into the front contours and was treated to the luxury of wind-up windows in the doors – a sharp contrast with the original Elite window which had to be removed if ventilation was needed!

The famous twin-overhead-camshaft engine came into existence as a result of a typically complicated Chapman deal whereby Ford would supply 1.5-litre five-bearing Cortina cylinder blocks and many other components, including the crankshafts, connecting rods and pistons, Lotus would design their own twin-cam conversion for the unit and J A Prestwich (JAP) of North London would machine it for them.

The original Elan was an open sports two-seater – and this is Michael Bowler, who was the long-time distinguished editor of *Thoroughbred & Classic Cars*, sampling a 1964 example.

The engine design and conversion was the work of Harry Mundy, then *Autocar*'s technical editor, already famous for his work at BRM and Coventry Climax – including designing the FPF Grand Prix engine along with Walter Hassan – and later to take charge of Jaguar's engine design team in the Seventies. The story goes that he had originally been commissioned to design an engine for Facel Vega, who then promptly ran out of money, and that the bare bones of his layout were revised and adopted for the Ford-Lotus application. Part of his brief was that the original Ford pistons were to be retained, which circumscribed the twin-cam valve gear somewhat, though he managed to keep the

included valve angle down to 54deg. The top-end was classic twin-cam, which is to say that it had two valves per cylinder and a part-spherical combustion chamber. Right from the start the 1.5-litre engine produced 100bhp in road-going form, but later racing units produced up to 180bhp with 1.8 litres and fuel injection.

It turns out that Colin Chapman offered Mundy either a flat fee of £200 for doing the job or a royalty of £1 per engine produced. Mundy, the realist, took the £200 – and regretted it ever afterwards. When Lotus built the 25,000th Lotus-Ford engine at the beginning of the Seventies and presented it to Ford with due ceremony, Mundy jokingly approached

After the open version, Lotus produced the Elan Coupe, a very civilized, if small, 'saloon' car. This was the S4 version, which had centre-lock steel disc wheels. Like the open model, many of these machines were sold as kit-cars to save the customer having to pay Purchase Tax.

The Elan Plus 2 was a definitive move upmarket, for the wheelbase was 12in longer than that of the Elan, the tracks 7in wider and there were two extra 'occasional' rear seats. The basic chassis layout, all-independent rear suspension and twin-cam engine were all retained. Lotus only sold the Plus 2 as a closed car ...

Chapman and offered to change his mind – Colin's reply was predictable!

Although the Elan predated the Seventies variety of Elite by 12 years, its layout had several important pointers for the future. Firstly, of course, it had the backbone chassis, all-independent suspension by coil springs and disc brakes at front and rear. It used a glassfibre body – eventually available in open or coupe form – and it featured proprietary items like a Ford gearbox and a Ford differential, though in a special Lotus casing. There were other proprietary components – instruments, tail-lamps, switches and many other details – all round the car, but over the next few years most of these would be replaced by special Lotus-designed items.

At this point the Lotus-Cortina project should be mentioned. Even though it was a very important car at Cheshunt in terms of numbers built, it was not truly a Lotus, but more a 'homologation special' built for Ford. At first it was a very specialized version of the Ford Cortina GT with standard bodywork and suspension. Lotus built these cars from 1963 to 1966, but the Mk 2 Lotus-Cortina which followed was to be built by Ford themselves, at Dagenham.

The Elan was a great success for Lotus in all respects except for that of reliability, and it could be supplied either as a built-up car or as a near-complete kit. The kit-car concept was purely British in origin, and had evolved because of a quirk in UK Purchase Tax laws. If a car was supplied complete, Purchase Tax had to be paid on the factory basic price, but if it was supplied as a kit and assembled by the owner and unpaid friends in a private garage, it came tax-free. There were all sorts of reasons why such procedures could be abused – and they were, frequently – but there was no doubt that this appealed to many enthusiasts. Until the beginning of the Seventies, the supply of Lotus kit-cars was a dominant part of the company's business.

Next, at the end of 1966, came the Lotus Europa – or Europe as it was known in, of all places, Europe – which was another example of Chapman's astuteness and co-operation with another manufacturer. Although the Europa was different from the Elan in almost every detail at first, it was still recognizably of the same design philosophy, even though it had a mid-engined configuration, and power and transmission by Renault. The Europa, in fact, used a different type of backbone frame to which the glassfibre bodyshell was permanently bonded at first, and the 1.5-litre Renault engine was an 82bhp unit adapted from that used in the front-wheel-drive Renault 16 hatchback.

In the meantime, other great events had been happening at Lotus. Not only had Jim Clark won two Formula 1 World Championships driving Lotus cars (in 1963 and 1965), but demand for the production cars had risen so much that the

... but Hexagon, one of their dealers, promoted this drophead conversion at one stage.

Cheshunt factory – which had been brand new in 1959 – was bursting at the seams, and there was no further scope for expansion. After a brisk search, Lotus found an ideal new site at Hethel airfield, near Wymondham, a few miles south west of Norwich. Not only was planning permission secured to build a new factory, but the airfield – an ex-USAF bomber base from the Second World War – had a ready-made runway and perimeter tracks, which would provide an excellent test track for the cars, and a landing strip for the Chairman's private aeroplane. As originally conceived, the factory was to have a 151,000sq ft single-storey production area and 26,000sq ft of ultra-modern open-plan offices and design facilities.

The target date for completion was originally set for October 1966, but there were some delays. An elegant but simple foundation stone in the office entrance now tells its own story: 'Laid, on the 17th of July 1966, by Colin Chapman, Founder.'

In fact the move to Hethel took place at the end of 1966, and in a very short time indeed Lotus had not only installed the production lines for their Elan (front-engined) and Europa (mid-engined) road cars, but they had also taken over the machine tools and jigs for production of the twin-cam cylinder heads; Villiers, who had recently taken over JAP, continued to assemble the engines for the first year or so.

Before long the total factory floor space had grown to

There was enough room in the rear seats of the Elan Plus 2 for small children, but if the front seats were pushed all the way back on their slides knee-room virtually disappeared.

The first mid-engined Lotus road car was the Renault-powered Europa announced at the end of 1966. It was a smoothly-shaped, but small car and strictly a two-seater.

The interior of the 1970 Lotus Europa, showing that familiar low roofline and the high central 'services' tunnel between the seats, which, of course, hides the deep backbone frame.

Open-plan office work at Hethel, soon after the new factory had been opened. The Europa is parked on the carpet near the front doors, and the directors' offices are in the background.

One of the first projects of Lotus' Eighties Managing Director, Mike Kimberley, was the Europa Twin-Cam, which involved mating the twin-cam Lotus-Ford engine to the Europa's structure. To improve rearward visibility for the driver, the 'sail' panels on the rear quarters of the shell were cut down.

The most powerful Europa derivative of all, regrettably only a one-off prototype, was GKN 47D, which was powered by nothing less than a 3.5-litre light-alloy Rover V8 engine. There hasn't been a V8-engined Lotus production car ... yet.

350,000sq ft – glassfibre body manufacture had originally been in old airfield outbuildings, but was soon brought into a new facility – and Colin Chapman had taken the opportunity to complete the full range of Sixties models, for in 1967 the Elan Plus 2 was revealed. This was a smooth fixed-head coupe model which effectively used a longer-wheelbase version of the two-seater Elan chassis, but was the first Lotus to offer 2+2 seating accommodation. The new model was more upmarket than the Elan, and it signalled the way in which Chapman wanted his company to develop. He wished to move away from the original Lotus image of producing proprietary-part kit-cars in cramped and rather crudely equipped factories – and he wanted to produce increasingly more prestigious machines.

Even so, the Elan Plus 2 was on offer as a kit-car as well as a fully assembled machine, though the proportion of fully assembled Plus 2s increased steadily in the next few years. It may be of interest to recall its principal features which, apart from the backbone chassis with all-independent coil-spring suspension, included front and rear disc brakes, rack-and-pinion steering and centre-lock pressed-steel wheels. There

was also the well-known Lotus-Ford twin-cam engine, tuned to produce up to 118bhp (net), and those truly elegant looks.

The Plus 2's wheelbase was 8ft, the unladen weight about 2,085lb and the original retail price for a built-up example was £1,923. Independent road tests soon showed that it had a top speed of 118mph and could often reach 30mpg (Imperial) in day-to-day motoring.

For the next few years company expansion was rapid. Until the late Sixties Lotus had been privately owned – almost entirely by Colin Chapman and his immediate family – but from October 1968 it was floated off as a public company. The original Board of Directors comprised Colin Chapman (as Chairman and Managing Director), Fred Bushell (Finance Director and Company Secretary) and Peter Kirwan-Taylor, who was then a power in the land with a merchant bank in the City of London. A few months later Dennis Austin was hired as Lotus' Managing Director. It is important, even in a marque history like this, to realize that the parent company was Group Lotus Car Companies Limited, and they controlled everything in which the resourceful Chapman had an interest, while Lotus Cars Limited was the wholly-owned

A display exhibit of the original Lotus-Ford twin-cam engine, with Harry Mundy-designed top end and Ford cylinder block and moving parts. The gearbox featured a Ford casing, but special close-ratio internals and – in this case – a very strange-looking lever.

subsidiary which was responsible for building the road cars. Then, as ever afterwards, the activities of Team Lotus were entirely separate from those of the road car concern.

It is important to the story of the Elite, Eclat and Esprit models to summarize how the company developed before they were born, and this can most simply be done with the aid of a table, which not only notes annual production, but also the pre-tax profits earned:

Production and profits – 1964–73

Year	1964	1965	1966	1967	1968	1969	1970	1971	1972	1973
Lotus production	1,195	1,234	1,519	1,985**	3,048	4,506***	3,373	2,682	2,996****	2,822*****
Lotus-Cortina production	567	1,118*	986	–	–	–	–	–	–	–
Total production	1,762	2,352	2,505	1,985	3,048	4,506	3,373	2,682	2,996	2,822
Pre-tax profits	£113,000	154,000	251,000	324,000	731,000	606,000	321,700	736,500	1,126,700	1,155,700

* This was the year in which the Lotus-Cortina achieved Group 1 homologation, which required 5,000 cars per year to be built.
** Europa production began early in the year and Elan Plus 2 production began during the summer.
*** The company's all-time record year. Europa sales began in the UK.
**** Plus 1,254 16-valve Type 907 engines delivered to Jensen Motors.
***** Plus 4,008 engines.

However, although profits had risen sharply in the early Seventies, production of Lotus cars had fallen away. There had been no new Lotus 'shape' since the Elan Plus 2 appeared in 1967 and no new chassis engineering either. Enthusiasts were beginning to lose interest in Lotus because they thought that the ideas had run out. Another problem for Lotus was that their twin-cam engine was built on the basis of the 1,499cc Ford cylinder block, which had been superseded in main-stream production by the deep-block (crossflow) 1,599cc design in 1967. Something would have to be done.

Something *was* being done – and the evidence was all around at Hethel. For the Seventies, Lotus not only had a new engine design in prospect, but new cars as well. It was going to be an exciting, if bumpy, ride.

18

CHAPTER 2

Lotus + Giugiaro = Esprit

Mid-engined flair, S1 and S2

In 1970, Tony Rudd, who had joined Lotus in 1969 and would soon be appointed Technical Director, had recommended the design of two major new Lotus models: the first, the M50, was the front-engined Elite described in Volume 1, and the second was the M70, a mid-engined machine intended to take over the mantle of the Europa. M70 was envisaged as a two-seater fixed-head coupe, with something of a wedge theme in its shape, and it was always intended to use as much as possible of the new running gear being developed for the M50/M52 Elite/Eclat cars. Like the M50 and M52, the new mid-engined M70 was given approval by management, though there was no way that design work on the new car could begin at once. Lotus' resources were still quite slender, and management first chose to concentrate on finalizing the Type 907 engine and the development of the front-engined cars.

The launch of a new design of mid-engined Lotus looked so remote that there was time for the Europa to be redesigned, not once but twice. Mike Kimberley was given the job of transforming the Europa, and in the autumn of 1971 the Renault-engined S2 was dropped in favour of the Europa Twin-Cam, a similar-chassised car powered by the 105bhp version of the Lotus-Ford twin-cam engine. This new car also featured a restyled rear bodyshell which offered better rear and three-quarter rear visibility, and it had cast-alloy road wheels, though the existing Renault four-speed gearbox was retained.

The new Twin-Cam – which was Type 74 in the Lotus scheme of things – had a top speed of 117mph, 0–60mph acceleration in 7sec and typical fuel consumption was about 25mpg (Imperial), a considerable improvement on the Renault-engined car. However, the Twin-Cam had a short life, for just one year later, in the autumn of 1972, it gave way to the Europa Special, which was effectively an uprated version of it, with the more powerful, 126bhp, 'Big-Valve' engine and a five-speed Renault gearbox. Performance was boosted yet again, this time to provide the little car with a top speed of 125mph and 0–60mph acceleration in 6.5sec. It set a very high standard indeed, and whatever was chosen to replace it would have a very difficult job to do.

In two distinct ways, however, it was going to be fairly easy to improve on the Europa – in the styling of a new car, and in the space offered to the passengers. The Europa, while always being immediately recognizable, had never been considered as an outstandingly attractive car, and even in its final developed guise it was most certainly not equipped with a very spacious cockpit. Like the two-seater Elan which preceded it, the Europa was a motoring machine rather than a passenger car. The M70, when it came along, would have to be more practical than this.

The styling of the new car, if not its engineering, began in 1971 following a chance meeting between Giorgetto Giugiaro and Colin Chapman at a motor show. Even in the early Seventies, Giugiaro had a formidable reputation as a stylist/designer, having started his career at Fiat, before moving on to Bertone, then Ghia, prior to setting up his own business, Ital Design, in 1968. Chapman knew all about Ital Design, and *everyone* knew about Lotus, so there was never any lack

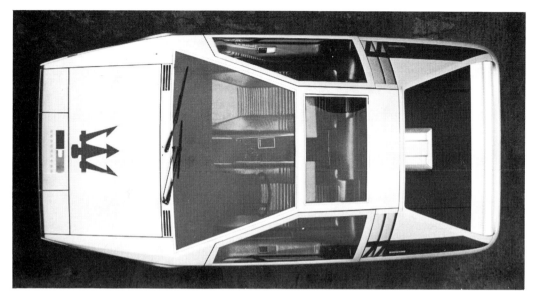

Perhaps this car – the Maserati Boomerang of 1972 – was one of the inspirations for the shaping of the Esprit by Giugiaro. Based on the running-gear of the mid-engined Maserati chassis, it was first seen at Geneva in March 1972.

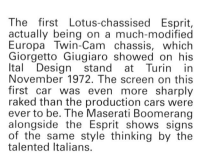

The first Lotus-chassised Esprit, actually being on a much-modified Europa Twin-Cam chassis, which Giorgetto Giugiaro showed on his Ital Design stand at Turin in November 1972. The screen on this first car was even more sharply raked than the production cars were ever to be. The Maserati Boomerang alongside the Esprit shows signs of the same style thinking by the talented Italians.

of understanding between the two. Quite simply, it seems, Giugiaro wanted to know if he could work up a special body style on a Lotus, and Chapman, with the M70 in mind, agreed to let him work on the basis of the mid-engined Europa.

Giugiaro had already produced the attractive mid-engined Bora for Maserati, and was working on the very angular, but startlingly advanced Boomerang project on the same chassis, so he was familiar with the challenges inherent in mid-engined layouts. To put it baldly, the very first Giugiaro style for Lotus was on the basis of a much-modified Europa Twin-Cam chassis, but since the Type 907 engine was soon to be installed, and the track and wheelbase dimensions were also altered, it is easy to see how Colin Chapman's mind was working. The Europa's wheelbase was 7ft 7in, and its widest track was 4ft 5.5in. Equivalent dimensions planned for the M70 – which did not have a name at this stage – were 8ft and 4ft 11.5in, respectively, so it was not surprising that the chassis supplied to Italy, thus lengthened and widened, was not Lotus' final word on the subject.

Work began on the style in mid-1971, and was completed

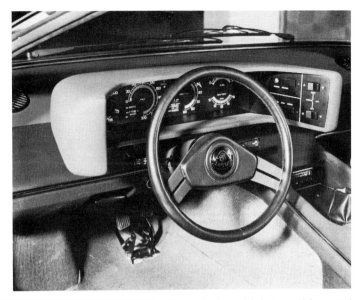

The interior, driving position and instrument layout – mainly mock-up – of the original Europa-based Esprit of 1972 with, would you believe, pedals which pivot on the floor? Nevertheless, an amazing amount of this concept was carried over to the production car of 1976.

Another view of the original silver Europa-based Esprit of 1972. There were, of course, many differences in detail style between this car and the definitive version. Apart from the rake of the screen, note the type of louvre behind the nose panel, the twin wipers and the way in which the rear of the body was arranged to open.

With neither help nor hindrance from Lotus, the British stylist Bill Towns – famous, among other things, for his work with Aston Martin-Lagonda – produced this comprehensive restyle of the Europa Special in 1975 and retained the centre-section, screen and doors. It was smart but unsensational, and was not adopted for production.

before the end of that year, not as a running car, but as a full-size mock-up in display trim. A second car, not only with doors which opened, but with a more advanced and integrated design of chassis, followed in 1972. It was the original silver-painted car – now remembered at Lotus, logically enough, as 'the Silver car'! – which made its public debut on the Ital Design stand at the Turin motor show of November 1972. Even at this stage, Giugiaro had dubbed it 'Esprit', defined by the *Concise Oxford Dictionary* as: 'sprightliness, wit', and students of styling evolution will want to be reminded that it stood alongside the Maserati Boomerang at that show.

At this time, it was interesting to see Colin Chapman explaining it away to the press as: '... an exercise on the basis of the Lotus Europa, to combine good styling with practical safety requirements ...', when almost simultaneously he was giving an interview to *Autocar*'s former editor Ray Hutton in which he commented: 'We should always have a model of an advanced sporting nature, such as a mid-engined two-seater.'

Something, for sure, was already on the move, and it was not long before motoring enthusiasts began to put two and two together. Mike Kimberley recalls that Lotus' reaction to

the completed prototype Esprit was so favourable that a design and development team was immediately set up to work with Giugiaro, and they stayed in Italy for at least 18 months. Chapman and Kimberley flew to Turin at least twice a week, during which the body style was refined and turned into a producible proposition.

After the tremendously favourable public showing of 1972 there was a considerable lull while mechanical design commenced, though in the Group Lotus company report published in mid-1973 one of the three pictures published under the heading 'The Coming Generation?' was of the Giugiaro prototype, which had already been adopted by the company. The first true production prototype was nearly completed by Christmas 1974, and was actually driven to London's Heathrow Airport to meet Colin Chapman when he returned from the Argentine Grand Prix in January 1975. Indeed, by this time, Lotus had confirmed that the Esprit would be launched during 1975.

The design and evolution of the new Type 907 engine is fully detailed in the companion volume covering the front-engined cars – Elite, Eclat, Excel and Elan – but it must be

What is a car like this doing in a Lotus book? The answer is that the Citroen SM coupe, complete with front engine and front-wheel drive, used the same basic transmission which has been fitted to every Esprit built since 1975. In the case of the SM, however, the gearbox was ahead of the line of the front wheels.

The definitive original-production Esprit of 1975, as revealed in this Lotus-prepared cutaway drawing, shows how the engine was placed ahead of the line of the rear wheels, but behind the passenger cabin, and the gearbox was behind the rear wheels and under the rather small luggage compartment. The car was arranged strictly as a two-seater, and there has never been a convertible Esprit production car – all those built having this wedge-styled coupe shell.

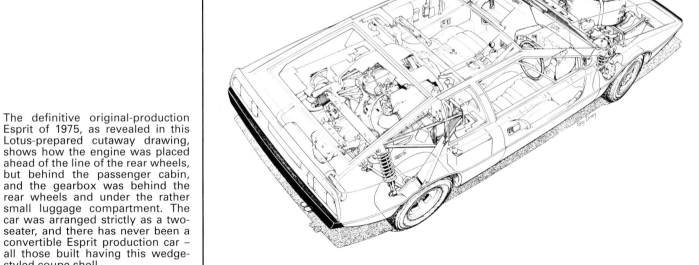

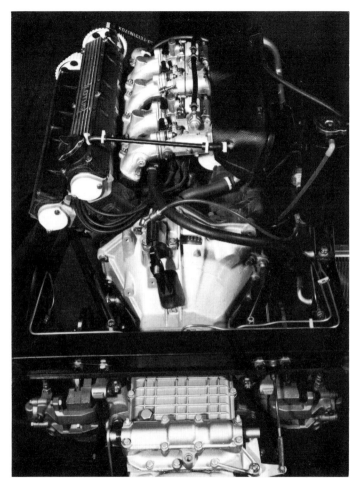

Brand new and sparkling – the engine/transmission installation of the Series 1 Esprit, showing how the gearbox and inboard disc brakes fit under the rear crossmember of the frame.

emphasized that it was always intended also to be used in Lotus' new mid-engined car.

It was, of course, an ideal package for the Esprit, compact in length, wide but not high, and – considering the power output – a very light unit. Both the cylinder block and the cylinder head were cast in aluminium, which was ideal for the Esprit, where there was bound to be a weight bias towards the tail.

For the original Esprit, therefore, Lotus specified a 2-litre version of the new 16-valve, twin-overhead-camshaft engine. As fitted in the Esprit's engine bay, behind the cabin, its installation and tune was exactly like that adopted for the front-engined cars. Complete with two twin-choke Dellorto DHLA carburettors, it was rated at 160bhp (DIN), and was installed with the cylinder block leaning over at 45deg towards the left side of the chassis.

Unavoidably this meant that there was a slight weight bias towards the left side of the car, but not even the most experienced testers could pick up any effect on the handling, so this detail was speedily forgotten.

To provide more interior passenger space and to allow for the use of the more bulky Type 907 engine, the wheelbase of the M70 was to be 8ft, or 5in longer than that of the Europa which it would replace. It was also destined to be a much wider car than the Europa, though it was always intended to feature a steel backbone chassis-frame.

Right from the start, Lotus' biggest problem was to find a suitable gearbox, and this was critical to the entire project. Since the Type 907 engine pushed out 140lb.ft of torque, even in 2-litre form – and the highest figure reached by the 'Big-Valve' Lotus-Ford twin-cams had been 113lb.ft – it was clear that the five-speed transmission from Renault, as used in the Europa Special, would not be strong enough for the job.

Chapman already knew that it was not financially viable for Lotus to design, tool and build their own transaxles – the five-speed gearbox for the Elite/Eclat cars used standard British Leyland gear clusters – for the cost of tooling up for cutting gears was immense, so Lotus had to look around for an off-the-shelf transaxle. At the same time, they had to consider the V8 engine project, which would produce a great deal more torque than the 2-litre 'four'.

One of the original Lotus press pictures of the Esprit of October 1975, before it was ready to go into production. This shot emphasizes the attractive wedge style, and it also shows off the separate front spoiler – which would not be blended into the styling before the S2 was launched.

Four big and powerful headlamps normally rest hidden behind flaps on the bonnet surface, but can be flicked up rapidly to a position where they ruin the aerodynamics!

The interior of the Esprit S1, which in concept is remarkably similar to that of the Giugiaro prototype of 1972.

Twin struts support the Esprit's rear hatch, which provides access to the luggage space and the engine, the latter normally being concealed beneath this moulded plastic cover.

Because Lotus were financially independent of any other motoring manufacturer, they could go shopping for a transaxle almost anywhere. But it was not as simple as that. They were looking for a five-speed transmission with not only ample reserves of strength, but one which satisfied their desire for mechanical 'elegance', was light enough and was guaranteed for continuity of supply for many years to come. With regard to the latter they were very wise, as the search for the transmission began in 1971, the first production units were not fitted until 1975–6, and they were still being used in the mid-Eighties.

The search eventually ended when Citroen offered Lotus the use of their five-speed all-synchromesh gearbox/final drive unit, which was being used not only in the exotic front-engine/front-drive SM coupe model, but also in the mid-engine/rear-drive Maserati Merak coupe.

The timing of the deal was important, for even in 1972 the SM was a young design at the peak of its popularity, and the

The Esprit S1 looked attractive from any angle. Roughly speaking, this would be the aspect of the car seen by truck and bus drivers – but not for very long!

The original design of Esprit chassis-frame – seen here in galvanized form – followed the famous Lotus backbone principle, though the lines did not seem to 'flow' as well as usual towards the rear, which is furthest away from the camera.

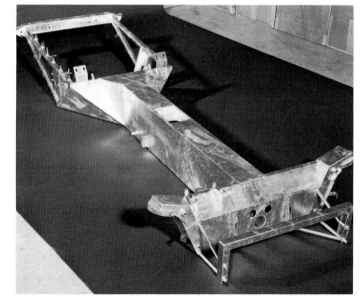

Maserati Merak had still not been announced. The SM transmission was a derivative of the five-speed gearboxes available on other large DS saloons and estates, and Citroen were able to offer supplies for at least the next 10 years. Even though the SM is now long dead and the Merak was dropped in the early Eighties, Lotus never had supply problems from France.

The gearbox was a conventional two-shaft design – conventional, that is, by transaxle standards – with the output of the spiral-bevel final drive from the second shaft. Its crownwheel and pinion design and the final drive casing were such that it could be run the 'right way' or 'wrong way' round – the Citroen SM and Merak installations, of course, work in opposing directions. A variety of internal ratios and final drive ratios could also be provided. In the end, those chosen for the mid-engined Lotus were the same as to be found in the SM Coupe, the original Merak and the later Merak 2000, but slightly different from those used in the more powerful Merak SS.

With the general layout of backbone frame chosen, the engine and transmission design finalized and the front suspension basically being the same as that fitted to the Opel Ascona/Vauxhall Cavalier, the rest of the mechanical design soon slotted into place. The independent rear suspension was as simple as possible; the fixed-length driveshafts doubled as upper transverse suspension links, combined coil spring/damper units were chosen, and large box-section semi-trailing radius arms helped to locate the wheels along with lower transverse links. Steering was by rack and pinion – but without power-assistance, no Esprit, not even the Turbo, ever having needed this – and the dual-circuit Girling brakes had front and rear discs, solid but not ventilated, with the rear discs mounted inboard. There was no servo assistance. Wheels were cast-alloy 14in diameter Wolfrace items, with 7in rims at the rear and 6in at the front.

Much work went into productionizing the startling Giugiaro shape, not only to make it easier and cheaper to build in quantity, but to make it meet all the regulations likely to face such a car in the mid-Seventies. The most significant change was to the angle of the windscreen. On the original 'Silver' prototype the screen had been angled back at a mere 19deg from horizontal, and to meet the regulations this had to be lifted to 24deg 5min. Colin Chapman, however, did not give in without a fight, and the production Esprit still kept the same dramatically swept screen pillars, a feature achieved by making the screen profile much less curved in plan than had originally been intended.

The interior layout and facia style were retained as much as possible and there was a great deal more space for two passengers, but no briefcases or other luggage could be stored in the wide cockpit. There was no space behind the seats, the cover over the backbone chassis-frame between the seats was high and wide, and there was only one storage container, ahead of the passenger's knees. As in the Europa, the seats were steeply reclined, and to climb in and out of the car was not for the modest or the unathletic.

In the meantime, there had been momentous changes at Lotus, both to the fortunes of the company and to the personalities at the top. Dennis Austin, Managing Director of Lotus Cars since 1969, moved on in 1974 and was replaced by Richard Morley, while Mike Kimberley, who had become Vehicle Engineering Manager in 1972, took over the title of Chief Engineer – from Tony Rudd – in 1974 and would be elevated to the Lotus Cars Board at the end of 1975. Tony Rudd became Group Research Director, a position he held until the early Eighties, when he was attracted back into Team Lotus and Grand Prix racing.

The Esprit was not ready for production when it was announced in October 1975, but for several good political reasons Lotus thought it necessary to reveal the car at the same time as the Eclat, which *was* ready. The car, after all, had already been around for three years by then and was known to the public; Lotus were worried that their customers would despair of it ever being announced if they did not show it then.

As has been explained [in Volume 1] the combination of energy crisis/oil shock, the launch of the M50 Elite and the progressive withdrawal of the Elan, Plus 2 and Europa families hit Lotus finances very hard. Pre-tax profits in 1973 had been £1,155,700, but they plunged to £293,909 in 1974 and losses were forecast for 1975. Faced with this sort of situation, the company *had* to retrench, and Lotus now confirm that the Esprit was delayed by about nine months due to this financial stringency. In normal circumstances, therefore, the Esprit would have been ready for deliveries to start on announcement in the autumn of 1975. The delay ensured that tooling was not complete by then, so that the first series production car was not commissioned until May 1976, and deliveries began in June and July.

It is worth recalling that the Esprit's UK price was fixed at £5,844 in October 1975 – when the comparable Elite 501 price was £6,493 – but this had rocketed to £7,883, representing an increase of 35 per cent by the time deliveries began. If Lotus had ever held hopes of producing a direct, but more upmarket replacement for the Europa Special, which had been dropped in 1975, they were now dashed. It is doubtful if the original published prices of October were realistic, for company cost accountants do not have to make such huge adjustments in a matter of months.

Few people would now argue with the opinion that the original Esprits were disappointing cars, for they were neither as fast, nor as refined or reliable as Lotus had hoped. For this, Lotus could certainly not blame their suppliers, for the

Engine assembly at Hethel towards the end of the 1970s, when production was at its height. Some are for front-engined cars, some for Esprit use. The engine in the mid-foreground already has the Citroen/Maserati transaxle fitted.

Esprit assembly at Hethel, with the rolling chassis already mated with the bodyshells, and with trimming and plumbing well under way. These are all S1 cars, with the separate front spoiler.

Lotus-built content of the car was approaching 70 per cent by value. The practical limits had already been reached, for the majority of the other 30 per cent went to pay for components such as the Citroen gearbox, wheels and tyres, electrical equipment, springs and dampers.

Lotus claimed that the original Esprit should have reached 138mph, but *Autocar*'s test car managed only 124mph, and *Motor* confined itself to a figure of 'more than 125mph.' This shortfall, however, was not as serious as the lack of refinement in the car, for much of the engine noise was transmitted to the cockpit, and the overall impression was one of harshness. The press, in general, thought that more than three years of development should have seen this ironed out before cars were delivered to customers.

Even so, no-one could argue with the car's remarkably sexy good looks, handling, general road behaviour and its overall effect on every other motorist – not least the gentlemen in blue! The use of early Esprits in films like the James Bond epic, *The Spy Who Loved Me*, where the car, or things mocked-up to look like the car, were made to perform incredible feats, must all have helped.

More than anything else, the Esprit was intended for sale in the United States. Peter Pulver, who was Lotus' principal Stateside distributor, ordered 150 cars at Earls Court in 1976, the first Federal Esprit was commissioned before the end of that year and deliveries began early in 1977. The Type 907 engine, complete with twin Zenith-Stromberg carburettors, came through the emission-reduction tests with such flying colours that the nett power output was still as high as 140bhp, so that the car had a top speed of 120mph in fourth or fifth gears (*Road & Track*, July 1977), all for an East Coast FOB price of $15,990 or – more importantly – $16,844 in California.

The USA launch had a dramatic effect on Esprit production, which had been 138 in 1976, but rocketed to 580 – the best Esprit year ever – in 1977. In 1976 all but four cars were built for the UK, but in 1977 no fewer that 474 were built for the USA.

Nevertheless, criticism and adverse press comment about the original Esprit had struck home, and Lotus made speedy attempts to improve the car. The result was that the S2 model was launched in August 1978, just over two years after the first Esprit deliveries had been made. Second thoughts, in this case, were wise ones, for the S2 was an altogether more integrated package.

Mechanically, there were few changes to the S2 compared with the S1, except that the 'E-camshaft' specification introduced on late-model S1s was now standardized – with a worthwhile improvement in mid-range torque – and Speedline road wheels replaced the original Wolfrace variety.

Externally, the most obvious improvement to the style was that the front undernose spoiler was now smoothly integrated into the shape of the car, while slightly protruding engine compartment air intakes were neatly positioned behind the

The 16-valve Type 907 Lotus engine tucked away in the engine compartment of the Esprit S1, before the top cover was fixed in place.

Perhaps this special Esprit was not quite as amphibious or versatile as the James Bond film *The Spy Who Loved Me* suggested, but it certainly made very good publicity for Lotus in 1976.

rear quarter-windows. At the rear of the car Rover SD1-type tail-lamp clusters were fitted.

Inside and underneath there was a twin electric motor lift mechanism for the headlamps, a new instrument cluster and slide-type switches, recontoured and wider seats, a digital clock, a redesigned engine cover and a revised aluminium-sprayed exhaust system. In case potential customers still couldn't tell the difference, there was an 'Esprit' decal on the nose instead of the 'Lotus' of the first cars and 'Esprit S2' decals on the rear quarters. The UK price, however, had rocketed once again, for the inflation rate was still quite shameful. In August 1978, therefore, the Esprit S2 was priced at £11,124 – 9 per cent higher than the last of the S1s. Lotus publicity chiefs were delighted to announce that Team Lotus' contracted drivers, Mario Andretti and Ronnie Peterson, had taken delivery of the first two production S2s.

There was one interesting mechanical innovation, made as much to suppress costs and to minimize the use of front underbonnet space as for good engineering reasons: the spare wheel now had a 5.5in wide rim and carried a small 185/70HR-13in tyre. It was a mere 'get you home' spare, not intended for prolonged use after a puncture.

Lotus, however, were not content with launching this new derivative, for they also produced a special 'Limited Edition' Esprit S2 at the first NEC motor show in October, which was decked out in black and gold 'JPS' livery to commemorate the Lotus 79/Mario Andretti feat of winning both Formula 1 World Championships; 100 of these cars were built, each individually numbered by a plaque on the dashboard, signed by Colin Chapman himself.

S2 performance was nearer to the original claims – Autocar's car was good for about 130mph, with 0–60mph acceleration in 8sec – and the drag coefficient of 0.34 was virtually unaffected by the changes to the style.

All in all, this was a step in the right direction, even if there were still advances to be made in refinement, reliability and overall creature comforts. Lotus had all these points in mind, and the launch of the Chrysler Sunbeam-Lotus in March 1979 hinted at the way they might move next. The Sunbeam-Lotus, after all, had an enlarged 2.2-litre engine. Would in-house Lotus models soon follow this development?

CHAPTER 3

Developing the breed

2.2-litre S2.2 and S3

The secret of further improvement to the Esprit series was to give it a more flexible and preferably more powerful engine. Lotus, however, was such a small concern that this could only be achieved by rationalizing engine developments among other models, *and* for Lotus' external clients.

Soon after Chrysler had approached Lotus to help them develop a new 'homologation special' for use in motorsport (see Appendix C), Lotus started work on an enlarged version of the engine, and it was always clear that this would eventually be used in updated Lotus models. This engine, known as the Type 911 when used in the new Chrysler Sunbeam-Lotus, would become the Type 912 when used in Lotus models.

Because Lotus engine designers had not built in much 'stretch' to the original Type 907, it was not practical to enlarge the cylinder bores as there was very little space between adjacent 'wet' liners. The only feasible way, therefore, was to lengthen the stroke, which was in any case welcome as this would provide the engine with more low-speed and mid-range torque to make the cars more flexible.

To make a 2.2-litre, the stroke was lengthened from 69.2mm to 76.2mm, and the capacity therefore rose from 1,973cc to 2,174cc. Although a new crankshaft was needed, many of the 2-litre engine's major components were carried forward for use in the 2.2-litre.

As far as Lotus were concerned, another significant feature was that the new Type 912 engine developed 140lb.ft of torque (the old Type 907's peak figure) at a mere 2,400rpm, which indicates just how much more punchy the larger-capacity unit had become.

This is a direct comparison between the three different engines:

Engine	Peak power at rpm	Peak torque at rpm
(Original Lotus unit)		
Type 907, 1,973cc	160bhp at 6,200rpm	140lb.ft at 4,900rpm
(Sunbeam-Lotus)		
Type 911, 2,174cc	150bhp at 5,750rpm	150lb.ft at 4,500rpm
(S2.2 Lotus unit)		
Type 912, 2,174cc	160bhp at 6,500rpm	160lb.ft at 5,000rpm

For a time, however, the demands of the DeLorean project, described in Chapter 8, meant that all other Lotus work had to mark time, and the bigger engine – a 2,174cc unit – was not ready until 1980.

At the same time as the front-engined S2.2 cars were announced, the Esprit S2.2 also made its bow. Since the Esprit had been updated in August 1978, the differences between S2 and S2.2 were limited to the use of a galvanized chassis treatment and the fitment of the Type 912 2.2-litre engine and a revised (partly stainless-steel) exhaust system. The price of the last S2 Esprit in Britain was £14,884, and that of the first S2.2 Esprit was £14,951 – a truly marginal increase. By the time it was withdrawn in favour of the S3 Esprit, less than a year later, that price had risen to £15,270. The S3 model, however, was a very different animal indeed.

The new features included in the S3 can best be summarized by quoting Lotus' own press material: 'The Esprit

As announced, the Series 2 Esprit had a facia and interior to this style, retaining the two-spoke Elite/Eclat-type wheel, and now with BL Princess-type slide switches in the pod.

Series 3 specification enabled us to rationalize a great many of the components, construction techniques and body/chassis tooling already incorporated in the Turbo Esprit ... (we now produce 76 per cent of the motor car at Hethel).'

Because the new chassis-frame and suspension systems were first seen in the Turbo Esprit, they are described fully in the next chapter. The S3 Esprit, launched in April 1981, used the same rigid backbone chassis-frame, revised rear suspension and modified front suspension, together with the larger disc brakes, as the Turbo Esprit, and it was also available with the 15in diameter Turbo Esprit wheels and tyres as an option.

The bodyshell modifications introduced for the Turbo Esprit — larger front and rear bumpers, the rear bumper with the word 'Lotus' embossed into it, and the new-style engine bay air intakes behind the rear quarter-windows — were also standardized, though the Turbo's spoilers and side skirts were *not* offered, even to customers waving wads of notes or dollar bills! A minor style change, which made much difference to the car's appearance, was that the lower sill and front spoiler mouldings were painted in body colour rather than in black.

Mechanically, the S3 used the same engine and transmission as the deposed S2.2, so the performance was unaffected.

There was also a 1978 'World Championship Commemorative' Esprit S2 with a Momo steering wheel and leather trim facings, plus other detail differences.

Drivers, however, would probably notice a different (Turbo-type) steering wheel and some trim and sound-insulation improvements.

More important than all these changes, welcome though they were, was the considerably reduced price of the Esprit. Here was a real bargain, for the S3 was not only a better car than the S2.2 it replaced, but cheaper as well. The S2.2's final price was £15,270, while that of the first S3 was a mere £13,461. That reduction of £1,809, or nearly 12 per cent, made many people sit up and take notice. It also did great things for the demand for Esprits, most of which were being sold in the UK: in 1980, 55 S2.2s had been produced for the UK market, while in 1981 there were 20 S2.2s and no fewer than 132 of the new S3s.

With the introduction of the S3 derivative, the Esprit 'came of age' and I can do no better than quote *Motor* in its test in August 1981: 'With a great many modifications aimed not only at reducing production costs but also at improving quality, it is a much better product all round, and a testament to Lotus' development abilities ...'

What was interesting about the Esprit S3 was not what it had, but rather what it did not have. For example, its drag coefficient was creditable at 0.33, but not sensational. As a Lotus engineer once told me: 'We never shouted about such things. In the early Seventies, even, we had Cd figures like that, and now other people are shouting about Cds of 0.35. We don't spend millions on advertising – we just build efficient cars.'

It is also quite remarkable that the 135mph Esprit S3 did not have ventilated disc brakes. Ridiculous, you may think, every high performance car has ventilated discs. But not Lotus. The engineers at Hethel looked long and hard at their braking requirements, especially when the 150mph-plus Turbo Esprit was being developed; they tested solid and ventilated disc brakes and concluded that the solid discs gave them better results over a longer period. In the past Lotus cars may have had quality and reliability shortcomings, but they could rarely be criticized on the grounds of engineering incompetence.

The Esprit S3 did not owe its splendid roadholding to exotic and expensive tyres such as the Pirelli P7, which are often found on Italian Supercars and can be extremely expensive to repair after a puncture. Instead, the S3 and the Turbo were equipped with Goodyear NCTs, which were more conventionally engineered, but still extremely grippy in all conditions – nor does it cost a small fortune to have a punctured NCT mended.

In 1983, Lotus announced the latest variation on the Esprit theme to coincide with their re-entry into the large US car market. Although the basis of the design was the successful S3 model, so much effort had gone into improvements to the bodyshell, the provision of more space inside the cockpit and tailoring the car to meet every US-market requirement, that the new Chairman, Fred Bushell, claimed that it was virtually another new model. Because of this, Lotus felt justified in calling the Federal car an S4.

However, during the Seventies, Lotus' involvement in the North American scene was not always happy or successful, and the entire range was withdrawn before any 2.2-litre-engined cars could be put on sale there. An explanation of this and an analysis of the problems Lotus now admit to have encountered, follows.

Export to the USA – the distribution saga

It is no exaggeration to suggest that Lotus' experience in selling cars to North America has often been disastrous. The United States is the richest potential car market in the world, recession or no recession, and Lotus' failure to establish a thriving market there with cars like the Eclat and Esprit was a real commercial tragedy. It is widely known that Lotus enjoyed a rush of sales in 1977 and 1978 and that sales collapsed immediately afterwards, but it is not as widely known why and how this happened.

On the face of it, any of the 16-valve-engined Lotus models should have found a ready market in North America. In engineering terms, there was no reason why the Lotus should be any less successful than cars like the mid-engined V6 and V8 Ferraris. In fact, for a short time it all looked very promising, but then everything seemed to go wrong. Why?

To summarize, I have prepared a table from figures provided by Lotus. Unfortunately, production statistics for Federal cars also include those destined for Japan, where similar specifications apply; however, I know that the last front-engined cars for the USA were built in August 1980, and the last Esprits in February 1980. That said, the table tells its own story:

Lotus production – Federal-specification* cars

Model	1974	1975	1976	1977	1978	1979	1980	1981	1982	Total
Elite S1	42	141	80	34	17	19	20	–	–	353
Eclat S1	–	3	106	125	68	17	20	–	–	339
Esprit S1	–	–	4	474	–	–	–	–	–	478
Esprit S2	–	–	–	–	251	128	13	8	–	400
Esprit Turbo	–	–	–	–	–	–	–	9	3	12
Total	42	144	190	633	336	164	53	17	3	1,582

* Includes cars for Japan

That charismatic moment which every factory visitor likes to see – the point at which a Lotus bodyshell is mated to the rolling chassis. This is a Series 2 Esprit on the simple production line at Hethel, now dismantled.

Compared with the original Esprit production car, the Esprit S2 had a different front spoiler and rear underbody profile, black plastic 'ears' behind the quarter-windows, new wheels, and – of course – a distinctive set of decals to announce itself.

There were air intakes at each side of the car on the Esprit S2, just behind the side windows, to help channel cold air into the engine bay. As with other Seventies Lotus road cars, there were fuel fillers on each side of the car.

The S2.2 Esprit replaced the S2 model in 1980 at the same time as the engine was introduced for the front-engined cars, and it was visually unchanged except for new badging. Under the skin, apart from the enlarged engine, the big step forward was the incorporation of the galvanized chassis-frame.

Spot the differences? Only the badging on the rear quarters gives away that this is the 2.2-litre-engined Esprit.

As the magazines used to say: 'The view most likely to be seen by other motorists' – the back of the Esprit before it accelerated away. This was the 1980 S2.2, complete with updated 'World Championship' badges on the engine lid.

John Lamm, writing in *Road & Track* in February 1983, had this to say: 'When Lotus updated its car line and image with the new Elite in 1974, and the Eclat and mid-engined Esprit a year later, it seemed the company was ready to continue offering the sort of automobiles we like: light, efficient, quick. And yet Lotus managed to fumble away what promised to be one of its best markets in the world, not because it lacked the right products, but because it couldn't build and distribute them for the US with anything like the brilliance that got the cars into production in the first place ... It's my suspicion that Lotus has just about used up all its credit with automotive enthusiasts in the US. It hasn't been an easy account to deplete.'

When North American sales of these cars began in 1974,

For the Esprit Series 3 of 1981, many minor but important improvements were made to the style, including the use of different sill shapes, an all-over body colour – instead of black sills – more wraparound bumpers, less prominent 'ears' and different wheels. The decals have changed, too!

On this Esprit S3 publicity car, the wheels are the optional BBS-type alloys normally fitted to the Turbo derivative. The word 'Lotus' was embossed into a revised back bumper.

The driving controls of the Series 3 Esprit were much as before, except that the two-spoke Momo wheel was now standard and the trim style had changed yet again. Like all Esprits, the handbrake of the Series 3 car lived in the door sill, just ahead of the seat.

The Type 912 engine, of 2.2 litres capacity, was fitted to all Lotus road cars by the summer of 1980. Although considerably changed internally, from the outside it looked much the same as before.

Lotus models were sold through a series of private distributors, with no centralized marketing and little forward planning. Lotus had great ambitions, however, and when one distributor suggested organizing national distribution to look after warranty, advertising, field service and sales, Lotus welcomed him, and Lotus Cars of America was set up.

But as Mike Kimberley told me: 'After 18 months, it was found that the organization had simply run out of money. In 1978–9 we were faced with seeing the market collapse by defaulting the customer on warranty costs, or we had to move in ourselves.

As the 1979 company report then said: 'At the commencement of the year, we terminated our previous arrangements ... and set up our distribution operation ..'

The new company was Lotus North America Inc, based at Costa Mesa, in California, with a total staff of *nine* people, but it soon became apparent that the operation was far too small to cover such a vast continent. More people and more money were required, and Lotus could provide neither. Between 1977 and 1979, the output of Federal-specification

Spot the differences? The engine with twin Dellorto carburettors is for UK or 'Rest of the World' use, while that fitted with twin Zenith-Strombergs is for the USA, or perhaps Japan.

cars mirrored the problem: 633 cars in 1977; 336 in 1978 and 164 in 1979.

It was at this juncture, when Lotus were agonizing over the imminent birth of the Esprit Turbo and the introduction of the 2.2-litre cars described above, that negotiations commenced with Rolls-Royce: 'David Plastow saw Colin and myself', Mike Kimberley recalls, 'and we came up with the idea of merging our US operations.' It was not so much a marketing merger as a takeover of Lotus' North American interests by Rolls-Royce, who already had an immensely well-organized, well-staffed business which had recently moved into new headquarters. A useful bonus was that this would work wonders for the 'Fleet Average' CAFE ratings of the Rolls-Royce and be a very low overhead for Lotus. This five-year deal was announced by both companies in September 1979, when it was pointed out that Rolls-Royce Inc had 68 USA dealers, all of whom would handle Lotus sales.

Almost at once, however, things began to go wrong. As Mike Kimberley says: 'Rolls-Royce took over the cars which we already had in North America and ordered another 125, then the car market in the USA died – and died in a big way. Rolls-Royce had their own problems because of that, for in 1979–80 the £/$ exchange rate moved against us by 42 per cent, which was appalling. The nett result was that our dollar prices had to go shooting up, orders disappeared and the distributors ordered no more cars from us. It's only fair to say that they had their own particular problems, with the Silver Spirit being announced – perhaps that was quite enough, without having to bother with Lotus.'

The result was that Lotus sales plummeted still further, and when the changeover from 2-litre to 2.2-litre cars was made, exports from Hethel dried up altogether. This explains why no S2.2 Elites, S2.2 Eclats, or Esprit S2.2, S3 and Turbo models were officially sent to the North American market after that.

Both companies realized that the tie-up had failed, and before the end of 1981 they began to negotiate closure and the unscrambling of a complicated situation; a public announcement followed in summer 1982, but all loose ends were not tied up until early 1983.

Financial upheavals at Lotus in the mid-Eighties
In the five years after Colin Chapman's untimely death, the Lotus company went through a corporate upheaval. Major shareholders came and went, crises were regularly faced – and overcome – and in the end the company was rescued from possible oblivion by General Motors. This ensured that changes and improvements made to the Esprit S3 in those years were often introduced with very little publicity. In that time, however, Lotus kept plugging away at the specification, the final cars being significantly faster than ever before.

Although Fred Bushell took over the chairmanship the day after Chapman's death, and despite Mike Kimberley and Alan Curtis both joining the board of Group Lotus at once, the concern was in deep trouble for a while. In the spring of 1983 Lotus were nearly forced to close, for the five-year loan of £2.2 million from American Express had reached maturity, and this was due to be repaid. There was neither cash in the bank nor sufficient cash flow for that to be done at once.

By August of that year, however, with David Wickins of British Car Auctions and Fred Bushell as the driving forces, a total of £6.69 million had been raised to secure the company's future. British Car Auctions subscribed £1.2 million in cash – and underwrote the issue of new shares worth £2.3 million – while Toyota injected £1.16 million, which gave the Japanese company 16.5 per cent of Group Lotus' share capital. The Chapman family, Fred Bushell and various trusts, still owned more than 20 per cent of the stock, with a mass of small shareholders holding the balance.

As Fred Bushell quipped at the Extraordinary General Meeting convened to approve this reshuffle: 'We were so nearly going under, you wouldn't believe it ...' Even so, in spite of this trauma, the company posted a £109,000 pre-tax profit for the first six months of the year, which compared well with a loss of £2 million for the whole of 1982.

When the dust had settled, it was seen that BCA either had, or influenced, no less than 47 per cent of the share capital, so it was no surprise to find that David Wickins was elected

Lotus engines were never simple, as this 2.2-litre Type 912 unit, in detoxed form, makes clear. But it could be much more complicated – there is no air conditioning pump or power-steering fitted to this unit!

Chairman in October 1983. Further changes followed in November 1984, when the construction machinery giant, JCB, bought an 11 per cent stake from the Chapman/Bushell holdings.

By the middle of 1985 there had been a further reshuffle, with Toyota increasing their stake yet again, and with the company renamed Group Lotus plc. Major holdings were:

BCA	29 per cent
Toyota	20 per cent
JCB	18 per cent
Schroeder Wagg (Merchant Bankers)	10 per cent
All other shareholdings	23 per cent

By this time, of course, the DeLorean scandal – which is

If Lotus were to improve the appeal of the Esprit for the Nineties, what were they to do? Designers like Peter Stevens saw that the style would have to be rounded-off, and the detail would have to become more integrated. Even so, many current customers were happy with the Giugiaro-styled wedge of the mid-Eighties. This was a 1986 normally-aspirated Esprit. The previous year Lotus had extended their chassis anti-corrosion warranty to eight years. Lotus specialists confirm that the galvanized chassis is virtually 'bombproof' unless damaged.

explained in more detail in Chapter 8 – had erupted, so there was never a time of what might be called financial calm and stability at Hethel.

The last truly major change then followed, in January 1986, when Wickins and his fellow directors realized that Lotus needed a broader capital base to allow them to expand to build the new-generation Elan, described in the companion volume. As a consequence, five large companies were invited to take over the business, but in the end it was General Motors of the USA, the world's largest and most diverse car-maker, which assumed control of Lotus, buying out every other shareholder. This transaction was formalized on January 22, 1986, and one immediate result was that David Wickins stepped down from the chair in favour of Alan Curtis, with Mike Kimberley continuing as Chief Executive.

Esprit S3 – the final improvements

In the traumatic final years just described, no visual changes were made to the Esprit, and on the surface it seemed that the normally-aspirated car, at least, was being neglected. Nevertheless, sales of the normally-aspirated model gradually rose, then fell away again. By this time most of these sales were being achieved at home, for the much more powerful Esprit Turbo had taken most sales in export markets.

In the autumn of 1984 Lotus announced that they were extending their anti-corrosion chassis warranty from six to eight years, but it was not until the end of 1986 that the HC (HC = High Compression) engine, already launched for Excel models, was made available.

The UK launch of the Esprit HC was actually delayed until early 1987, when it was seen that the latest version of the 2.2-litre engine, dubbed 912S, had a compression ratio of 10.9:1. Along with its new Nikasil cylinder liners and an improved cooling system, this allowed peak power to be boosted to 172bhp.

In the Esprit HC, peak power and torque figures were both slightly lower than in the Excel, this being because there was a slightly more restrictive exhaust system in the mid-engined Esprit. Other improvements over the old S3 (which was discontinued) were the use of adjustable-rake seats, a larger-capacity battery and full engine and transmission undertrays to improve the aerodynamics. With the car thus modified, Lotus raised the UK retail price from £18,980 to £19,590 – an increase of £610, or just 3 per cent.

This, however, was merely a temporary improvement, for the existing-shape Esprit only had another few months to live. In the autumn of 1987 the old car was discontinued, and a new 'X180' style of Esprit took over.

Except in detail, the mid-Eighties Esprit S3 facia/instrument layout was much like it had been in 1976 when the car was originally put on sale. One interesting – and none-too-popular – detail of this 1986 cabin was the post-Chapman badge on the two-spoke steering wheel.

CHAPTER 4

Turbo Esprit

The 2.2-litre supercar

Make no mistake, the Turbo Esprit was not just a slightly modified Esprit, nor just an easy and conventional way of gaining incremental sales for Lotus. In the Hethel scheme of things it was very important and technically very significant, if only for the fact that it was chosen to spearhead a new assault on the United States market in 1983.

It is tempting to suggest that if only Lotus had been able to give the car a different body style, they would certainly have been justified in giving it an entirely different name as well. Under the skin, which was extensively retouched by Giugiaro, there was a new chassis-frame, new rear suspension, new aerodynamic features and a turbocharged version of the 2.2-litre 16-valve engine, which produced no less than 210bhp. For this amazing car, Lotus claimed a top speed of 152mph – and it meant that they had produced their first true Supercar.

By any engineering standards, the Turbo Esprit was, and is, a phenomenal motor car. However, like the original Esprit, it was first shown to the public a long time before deliveries could possibly begin. The occasion of the car's launch was an extravagant party at the Albert Hall, in London, hosted by Team Lotus' Grand Prix sponsors at that time, Essex Petroleum, in which one of the three prototypes was displayed in the dramatic Essex blue, silver and red livery.

Although the Turbo Esprit was the first Lotus actually to be exhibited with a 2.2-litre engine, the normally-aspirated version of this unit had already gone into production for the S2.2 models. Even so, because of the rush to show a Turbo Esprit at the Albert Hall, the planned release of non-turbocharged engines was overshadowed for several months.

Lotus' own press material stresses the scope and nature of the new car's development: '... this new addition to our model range is not an Esprit with a bolt-on Turbo pack, but a fully developed and redesigned motor car in its own right.' So much of the car was new, indeed, that it would probably be easier to list what was *not* changed, modified or improved. For a start, there was a new design of backbone frame, prepared not only because Lotus wanted to provide an altogether more integrated structure than before, and to accommodate a new rear suspension, but also because they wanted to leave enough space for the still-secret V8 unit to be fitted one day. Lotus made no secret of the existence of a large-capacity V8 engine in their development programme, which explains why the engine bay of the new frame was wider and the general layout so much more sturdy.

At the front of the car, the independent suspension – now with more Lotus-sourced parts than ever – and rack-and-pinion steering of existing Esprits was retained, but at the rear there was a new layout. Earlier cars had used the simplest possible linkage, in which the fixed-length driveshafts doubled as upper transverse links. The disadvantage of this was that cornering stresses were fed into the final drive housing, found their way to the rest of the engine/transmission assembly and did little to minimize harshness and vibration in the structure. For the new car, there was a new linkage, with a wide-based lower wishbone and an upper transverse link, which allowed the driveshaft to have sliding joints and to carry out only one function; coil spring/damper units, of course, were also retained.

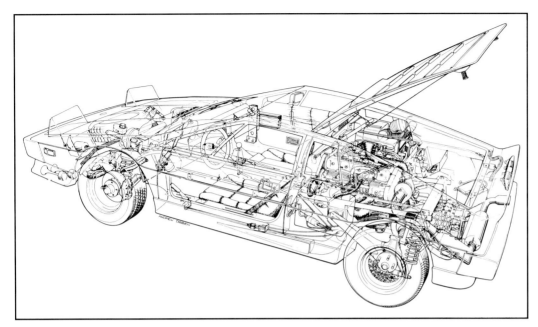

Andrew Dibben's excellent cutaway drawing of the Esprit Turbo, in left-hand-drive form, shows off the neat packaging of this exciting mid-engined model. There is not an ounce of wasted weight, or an inch of wasted space.

The schematic installation of the turbocharger to the 2.2-litre Lotus engine. In fact the turbocharger itself is not on top of the engine, but immediately behind it, above the clutch bellhousing.

Development of the new car began before the end of 1977, and if so much effort had not needed to be diverted into the DeLorean project, it would certainly have been announced months earlier than was the case. Even so, the M72, as the Turbo Esprit was known – M71 was the project including the V8 engine, by the way – progressed from 'good idea' to production car in little over two years. Apart from the new chassis and suspension, early decisions had to be made about the body style and the changes to the engine.

Compared with the bodyshell on normal Esprits of the period, there were many obvious changes and additions, mostly made for good aerodynamic reasons. The main shape and structure was unchanged, but differences were obvious from all angles. At the front, there was a larger wrap-round bumper allied to a deep new spoiler. There were matching skirts along the sides, under the doors, complete with NACA-type ducts moulded in to direct cooling air towards the engine compartment.

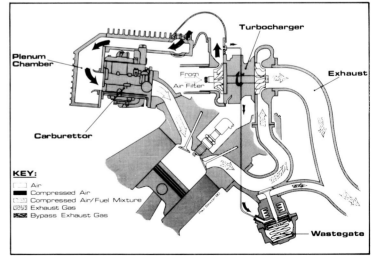

Plenum Chamber

Carburettor

From Air Filter

Turbocharger

Exhaust

Wastegate

KEY:
☐ Air
■ Compressed Air
▨ Compressed Air/Fuel Mixture
▨ Exhaust Gas
▨ Bypass Exhaust Gas

The complete Esprit Turbo engine/transmission assembly, showing the complex cast exhaust system leading to the turbocharger, which is mounted above the clutch bellhousing, and the inlet trunking at the top of the unit. The Citroen SM-type gearbox is unchanged from the normal Esprit installation. The all-important wastegate control is low down under the exhaust trunking, close to the turbocharger.

The carburation side of the Esprit Turbo engine/transmission shows virtually no evidence of turbocharging at all, except that the intake trunking to the carburettors is obviously different – and the word 'Turbo' is cast into the ribbed passage.

At the rear there was a sizeable and completely functional spoiler above the tail-lamps and number-plate, a large bumper matching the front component and extra skirting under the tail. Above the engine bay, instead of glass there was Venetian-blind-style black louvring. In addition, not only to make the styling even more distinctive, but to optimize the roadholding, there were 15in diameter alloy road wheels of a new style, with 7in front and 8in rear rims and Goodyear NCT tyres. Most Turbos have been sold on BB's wheels, but a few early cars were supplied with three-piece Compomotive rims instead.

To produce the new Type 910 engine Lotus redesigned the Type 907 from end to end, and the final product differs in almost every detail from normally-aspirated Type 911 (Talbot Sunbeam-Lotus) and Type 912 (S2.2 Lotus) units. The finalized engine had its Garrett AiResearch turbocharger mounted above the clutch bellhousing, behind the cylinder block, and boosted inlet air to the Dellorto carburettors to a peak of 8psi above atmospheric pressure. To compensate for this, and to optimize all settings, the nominal compression was reduced to 7.5:1 (from 9.4:1) and there was a different camshaft profile. A feature not always noticed by the pundits was the dry-sump lubrication system.

The result of a great deal of development work was not only a very powerful engine – peak power was 210bhp (DIN) at 6,250rpm and peak torque 200lb.ft at 4,500rpm – but one which was extremely flexible, having mostly 'non-turbo' habits. Its behaviour was so unexpectedly good, for instance, that *Motor Sport* headlined its test, 'The perfect driving machine?', while *Autocar* called the very same car a 'paragon of the turbocharged'. To deal with all this power, the only important changes were to the clutch – whose diameter was increased by an inch – and the brakes, which were larger – but not ventilated – at front and rear.

Inside the car, it was still the same basic layout as before, with no more space for stowing oddments, cases or especially children(!) though there was a new and plushy type of trim and padding, while the Essex cars were given air conditioning as standard and a complex Panasonic radio cassette player

When Renault racing engine designers visited Lotus to finalize their GP contract they looked around the factory. One of their designers inspected the Esprit Turbo engine, looked at the ribbed cast-alloy air passage from the turbocharger to the carburettors, and exclaimed: 'Ah yes, your inter-cooler!'. Not specifically intended to be so, the passage nevertheless has intercooling properties.

The rolling chassis of the Esprit Turbo, showing that the 210bhp engine takes up virtually no more space than the ordinary unit.

When covered by bodywork, the Esprit Turbo presents a bonnetful of engine to the enthusiast. On this model only special camshaft covers are used so that the 'Lotus' name is facing directly upwards.

mounted in the centre of the roof panel; unfortunately this was FM (VHF wavelength) only, so was disliked by most customers and was dropped when the Essex Commemorative run was complete.

Although the Essex Turbo Esprit, price £20,950, had been announced in February 1980, deliveries did not begin until August that year, immediately after the summer holiday shutdown. However, although development work had been completed to 'Federalize' the sensational engine, the Turbo Esprit, as already explained, was not sent to the United States at this stage. In its first calendar year, 1980, therefore, a total of 57 Turbos were built, of which 44 stayed in the UK.

Apart from the price of the Turbo – which in January 1981 had not changed from the launch figure – another feature which seemed to disturb some potential customers was the flamboyant Essex colour scheme. Very wisely, therefore, Lotus decided to offer a more conventional machine once the intended 100 Essex cars had been built. From April 1981 a Turbo Esprit in conventional colours, but still with all the

This much-modified galvanized chassis-frame was first seen on the Esprit Turbo but was later to be standardized on all Esprits. The extra space around the engine bay was provided not only to accommodate the turbocharged 2.2-litre engine, but also to make way for a 4-litre four-cam V8 which Lotus subsequently admitted they were developing.

Well, it makes a good picture, even if the advertising connection is plain. The car was as prepared for the James Bond film *For Your Eyes Only*, and Goodyear's airship *Europa* speaks for itself. At that time, Lotus F1 cars were running on Goodyear tyres.

appropriate body panels and decals, was made available for £16,917, though part of this huge price reduction was due to the fact that air conditioning was now an optional extra. Demand perked up almost at once, and in spite of the generally reduced level of Lotus sales in 1981 and 1982, Turbo production rose to 116 in 1981 and 205 in 1982. In 1982, the Turbo was the fastest-selling Lotus model, backed up by the normally-aspirated Esprit S3, of which 160 cars were built. The glamorous publicity occasioned by the use of Turbo Esprits in the James Bond film *For Your Eyes Only* did no harm at all!

Lotus had had several traumatic experiences in trying to sell and support cars in the United States, but Mike Kimberley thought that a new deal would offer the best of all possible worlds. Before the end of 1982, a new company,

Lotus Performance Cars, was set up on the East Coast, headed by John A Spiech, previously General Manager of Ferrari, North America – a 1,000 cars per year operation – and it was suggested that 350 cars would be sold in 1983, with 700 projected for 1986.

As Mike Kimberley told me: 'We have a special car for this market, which is really a brand new model, with a new set of body moulds, a lot more space, foot boxes and headroom, and so on. We're going in for the first time ever with a clean market. There hasn't been mid-engined Lotus for sale over there since 1981, and the Turbo was never launched over there, so it's all new.

Even though the 1983 Federal Turbo Esprit kept its original project number of M72, it was new or modified in many respects, and factory personnel were very excited about

The Esprit Turbo, complete with Essex Petroleum colour scheme, made its bow early in 1980. The basic Esprit style was embellished with a deeper front spoiler and skirts along the sides, and the wheels were 15in diameter, with wider rims. Another recognition point was the slatting over the engine.

This side view of the 1980 Essex Esprit Turbo – of which 100 examples were made – emphasizes the extra aerodynamic features of the bodyshell, including the deeper front spoiler, the larger rear spoiler and the F1-type skirting along the sides. The NACA ducts immediately below and behind the doors help to direct extra air into the hot engine bay.

The squashy-looking interior of an early Esprit Turbo, with the basic instrument and control layout unchanged.

This complex-looking communications panel by Panasonic was only fitted to the Essex Esprit Turbos. It was attached to the inside of the roof panel, high above the transmission/services tunnel.

its prospects, especially as the peak power output had needed only minor trimming. Lotus viewed the realignment of £/$ exchange rates as an unqualified bonus for them as exporters. Early in 1983 they were expecting to price their Turbo Esprit at $48,000, compared with the $60,000 asked for the equivalent two-seater Ferrari 308. 'And,' as Kimberley quoted with glee, 'our car will do 0–60mph in 6.5sec in full detoxed trim, whereas that Ferrari takes 8.3sec.'

After March 1983, all Turbo Esprits were built with conventional wet-sump engine lubrication. The dry-sumping introduced in 1980 was always agreed to be a real safety-conscious 'belt and braces' job, and prolonged testing convinced the engineers that it was not really needed.

The success of this car, and the almost universal praise for its engineering, behaviour and detail fittings, all go to prove that at this price a customer is more interested in the right specification than in mundane practicality. If he is not in the Turbo Esprit price class, he simply would not appreciate that the extrovert good looks also have a functional effect on the car's aerodynamics, but he would appreciate the intercooling

The front stowage compartment of the Esprit Turbo really has no space to accommodate anything except the proverbial toothbrush. The wheel is a small-diameter/speed-derated spare – merely a 'get you home' or 'get you to a garage' emergency fitting.

The Esprit Turbo facia layout, virtually the same as for the normal Esprit. The turbocharger boost gauge is in the centre of the instrument board between the speedometer and rev-counter, the layout (in 1980) being completely different from that of the contemporary S2 Esprit.

The Esprit Turbo looked just as eye-catching without its Essex livery. This white example was parked in the centre of Norwich for a photocall. Note the 'James Bond' number-plate!

Prince Michael of Kent is a real motoring enthusiast, and he thoroughly enjoyed a demonstration drive in a JPS Esprit Turbo at Donington Park, with Nigel Mansell at the wheel.

effect of the finned manifolding from turbocharger to carburettor plenum chamber, the remarkable grip and handling and the excellent lie-down driving position. And if he were a Turbo Esprit enthusiast, he would make sure that he never went kerb-hopping to cause punctures anyway.

To avoid confusion in future years, the existence of the Bell & Colvill Esprit Turbo must be mentioned at this stage. Bell & Colvill, based at West Horsley, in Surrey, are Lotus dealers, and in 1978 they announced their own privately-financed turbocharged conversion on the basis of the Esprit S2. This car, of course, was the 2-litre model, and the turbocharging layout was entirely different, having been developed by Stuart Mathieson on his own account. In this conversion, priced at £2,000, there was a single and very large SU carburettor upstream of the Garrett AiResearch turbocharger, which fed the fuel-air mixture at a maximum

Not to be confused with the factory machine, the Esprit Turbo produced by Bell & Colvill, the well-known British Lotus dealers from Surrey, had entirely different carburation and trunking arrangements as a conversion of normally-aspirated Esprits.

For 1984 the Esprit Turbo was improved in detail, notably with an enlarged boot compartment, and with the interesting option of a removable glass-panel sunroof. Not much use at 130mph, you might say, but in certain climates and restricted speed conditions (California, or southern Europe perhaps?) Lotus thought it would be very popular.

boost of 8psi to the engine, whose nominal compression ratio had been reduced to 7.5:1 by the use of new pistons. Peak power was quoted at 210bhp at 6,000rpm, while peak torque was 202lb.ft.

Although it was neither as flexible, nor as refined as the factory Turbo which was to follow, the Bell & Colvill car was undoubtedly very fast, with a claimed top speed of more than 150mph, 0–60mph in 6.2sec and 0–100mph in 17.2sec. The only way that one could identify this car externally was because B & C had added 'turbo' decals to the front and sides of the car, near the factory's own 'Esprit' decals.

From this angle it was impossible to see the optional glass-panel roof in the mid-Eighties Esprit Turbo. Unhappily, as far as many Lotus enthusiasts were concerned, it *was* possible to see the new type of bonnet badge. This did not survive to the end of the Eighties. From this viewpoint, too, the low-mounted air intakes for the engine bay – ahead of the rear wheels – identify an Esprit Turbo from the normally-aspirated model.

Compared with the normally-aspirated Esprit of the mid-Eighties period, the Esprit Turbo's facia layout was nearly identical. Note, however, the turbo boost gauge, above and between the speedometer and rev-counter dials, and the different badging on the centre console.

Esprit Turbo in the mid-Eighties

From 1983 to 1986 the Esprit Turbo was virtually buried under the tide of controversy which surrounded Lotus, their finances and their many commercial problems. Somehow, though, customers rose above all this and ordered the car in increased quantities. The Esprit Turbo was Lotus' best seller in each of those years – and in much-modified early-Nineties form it was the only car which kept Lotus afloat. The corporate upheavals have already been summarized in Chapter 3.

In the autumn of 1983 the Turbo was modified by being given a rather more spacious boot compartment, but there was also the introduction of an intriguing option, a removable glass panel in the roof. It made a good car even more versatile. In 1983, 343 cars were built and no fewer than 418 would follow in 1984. In all this success, the building of Lotus' 30,000th car, in May 1984, passed almost unnoticed.

Except that the new eight-year anti-corrosion warranty was applied to the Turbo's chassis from late 1984, there were to be no more significant changes until March 1986. That was the year in which not one, but two powerful new derivatives were launched: one for sale to the USA and other 'strict-

By 1986 the Esprit Turbo was coming close to the end of its seven-year career. Here, posed ahead of a sand dune (that must be the highest hill in Norfolk, by the way ...) the 1986 model looks very similar to its sister type of the early Eighties. As ever, the louvred cover behind the cabin is a recognition point – and that alien badge.

Except that one usually recognized the Turbo HC model of 1986–7 by its increased performance, Lotus also provided different decals on the flanks, immediately behind the door pillar.

Before the end of its run in 1987 the Esprit Turbo had been given seats with adjustable backrests but this was little use to a tall driver who pushed his seat all the way to the bulkhead.

emission' markets; the other for sale to the rest of the world.

In March 1986, at the Geneva motor show, the Turbo HCPI (HC = High Compression, with Petrol Injection) was unveiled. The compression ratio had been increased to 8.0:1, maximum boost pressure was up, and not only was this the first Lotus to use injection – the familiar Bosch K-Jetronic layout – but it also had a catalytic converter in the exhaust system. The packaging engineers had done their best to make the cockpit more roomy, for the seats had been widened and lowered, while the footwell area had been enlarged.

Not only was the engine more powerful *and* more torquey than before, but the chassis had been improved with wider-section tyres (195/60s at the front, 235/60s at the rear), while there was a new front spoiler and a larger radiator intake. To drill home the message this car also had HCPI decals.

The 'Rest of the World' derivative of this car, titled Esprit Turbo HC, was put on sale in October 1986. It shared the same high-compression head of the HCPI, but retained its Dellorto carburettors and was not fitted with a catalyst. Like changes made to the normally-aspirated Esprit at the same time, the latest HC also had an uprated cooling system and adjustable-rake seats. It was a more expensive car than before – £24,980 in the UK, which was an increase of £1,540 on the original type.

Compared with the earlier Turbo there had been a 10 per cent torque increase, which made an immediate and obvious difference to the performance. When *Autocar* tested the car in 1987, it was summarized as: 'In many areas ... a very practical supercar ... a remarkably satisfying device with which to cover long distances quickly. But with many less expensive sportscars now offering similar performance, the Lotus begins to look a little less attractive than it did three years ago. The same cannot be said for its styling, however, which remains truly exotic.'

By that time, however, the original-shape Esprit was only a few months away from the end of its long career, though Lotus managed to keep this secret well hidden until the very last month. In the autumn of 1987 the old car was discontinued and a new-style, more rounded, X180 model took over.

CHAPTER 5

New-style Esprits

Major changes for 1988

In 1987 everything looked rosy for Lotus. The first prototypes of a new smaller front-engined car – to be called Elan when launched – had gone on the road, Group Lotus had made more than £2 million profit for their new masters, GM – and a restyled Esprit was announced.

The new-style car, coded X180 at Hethel, was launched in October 1987, with deliveries beginning almost at once. Because this was merely a new bodyshell on an existing and well-proven chassis, with a new gearbox/transaxle, this was an excellent, low-investment way of developing a new model. The progress from concept to production had taken only 15 months.

The problem – if it *was* a problem – was that in spite of its more rounded character, the new car looked so very much like the old. Compared with the now-obsolete and sharp-edged Giugiaro style, the X180 was more rounded, smoother and softer than before. Its proportions were virtually the same, but it was as if the hot sun had been allowed to play for hours on a rather pliable model stack. One easy way to 'pick' the new design was by its use of a brand new style of cast alloy road wheel.

The aerodynamic impression of shapes, incidentally, can be deceptive. Although the new car looked much smoother than before, its drag coefficient was in fact slightly higher, at 0.35 instead of 0.34.

Colin Spooner's team of designers, led by Peter Stevens, had solved a near-impossible task with great style – literally. The aggressively sharp-edged Giugiaro design which had been truly in vogue in the Seventies but had been outdated by design trends in the Eighties, was replaced by something more sensuous, more gently rounded and more sophisticated – yet it sat on the same backbone chassis, with the same wheelbase and track dimensions.

As with other modern Lotuses, the new style was designed to be produced in two large halves – top and bottom – mainly from glassfibre, but with some local Kevlar reinforcement, using the company's patented vacuum-assisted resin-injection (VARI) process. The style was created in-house, at Hethel. Giorgetto Giugiaro, I understand, was not asked to offer ideas, and probably never even saw the new shape until it was unveiled.

Although the profile of the new body varied by no more than an inch from the old at any point, it looked very different. All key lines were rounded off, rather than razor-sharp, most details – such as the fuel-filler flap, and the air scoops for the engine bay, which were in the sill mouldings ahead of the rear wheels – were much tidier than before, front and rear bumpers were made in knock-resistant mouldings, and – in spite of initial impressions – every single pane of glass was a new shape and size. The new body featured a lift-out panel in the roof, this being either in a Nomex honeycomb material, or in tinted glass.

There were several basic visual differences between the normally-aspirated and the turbocharged types. Each car had its own special type of cast alloy road wheel – though the same tyres were shared between types – while the Turbo's front end had extra driving lamps and a different front-end air intake. The biggest and most obvious difference was clear

In October 1987 Lotus dropped the original Giugiaro-styled Esprit, replacing it by a newly-shaped Esprit whose style had been penned by Peter Stevens. The new car was much more rounded and every moulding and piece of glass had been changed, yet its overall profile was very similar indeed. This was the turbocharged version – there are badges behind the door shut line.

from the three-quarter-rear aspect. Both types of car featured neatly detailed 'flying buttress' panels from the rear of the doors to the flip-up spoiler on the tail, but only the turbocharged car filled in most of that recess with a large sheet of glass to give what Lotus claimed was a 'tunnel-back' feel. The Turbo also had a different rear lower body, with a carefully-profiled under-tray/spoiler to act as a scoop to help extract hot air from the engine bay and the brakes.

Inside the car there was significantly more legroom and passenger space in general – but this was still a small two-seater coupe cabin because there was a limit to what Grand Prix designer John Cooper would call the 'Albert Hall' effect which could be achieved. Instruments and the panel were new, there was a better ventilation system – which was sorely needed, as all hot-climate Esprit owners would no doubt agree – yet this was still not a car in which tall, particularly well-built drivers could get comfortable.

Facelifts, in general, do not always work very well, but the X180 process was in any case much more than a mere facelift. It was a complete restatement of the Lotus mid-engined theme, and it worked out extremely well. The designers had worked round the car in great detail, producing a neater and more integrated solution to almost every task.

The new car looked so good that it was a temptation to go

Compare these two side-on studies, and you should be able to 'pick' one X180-styled Esprit from the other, though the differences are small. The car above is the Turbo version, complete with 'Esprit Turbo' badges behind the doors, while below is the normally-aspirated model, which carries its Esprit badges below the rear quarter-window. Note the 'Lotus design' badge ahead of the door pillars.

Two more 'spot the difference' studies of the X180 Esprit. This is the normally-aspirated model, as introduced in the autumn of 1987, with what was known as the 'open back' body style between the two sail panels on the rear quarters ...

... while this was the 1988-model Esprit Turbo, which had a 'glass-back' feature between the sail panels, and a different rear lower body moulding and cooling grille arrangements.

The Esprit in its original S1 form in 1976. The car was intended to satisfy the two-seater Lotus market previously served by the mid-engined Europa.

Like the last versions of the Giugiaro-styled Esprit of 1975–87, the X180 Esprit had reclining seat backs. However, as can be seen here, most of the time the seats would be pushed back hard against the bulkhead, giving no scope for the backrests to be reclined. Note the fixings for the glass roof panel.

back to the old type, criticize slots, flaps, lines and features and say to oneself: 'How could we ever have put up with *that*?'. It wasn't heresy, but of how many other Giugiaro designs would one even begin to be critical?

Though little could be done about the overall size of the two-seater cockpit – particularly as no changes had been made to the central backbone, or to the height of the chassis pressings – the view from the driver's seat was little different from before. There was a new and more integrated facia design dominated by VDO instruments, the roof panel could be removed – though surprisingly few owners seemed to take advantage of this feature – and improvements had been made to the ventilation system.

Mechanically the new-shape Esprits were very similar to those of the 1986–7 HC models which had just been made obsolete which is to say that for normally-aspirated cars the 172bhp engine was standardized, while for the Turbo there were two different types, one with Dellorto carburettors, the other with Bosch K-Jetronic fuel injection. There was one major exception – but only on non-Federal-market cars at first. Instead of the familiar Citroen/Maserati type of five-speed gearbox, Lotus had to find a replacement that would fit, which was robust enough and suitably packaged.

Because the number of powerful European cars with combined gearbox/final drive transaxles was increasing all the time, Lotus had no trouble in finding a new supply. Links with Renault of France, which had flourished in the Sixties in the case of the mid-engined Europa, were revived, and Lotus chose the robust Renault 25/Renault GTA type of five-speeder instead of the old Citroen type. Surprisingly, though, the Citroen gearbox was retained on USA-market Esprit Turbo HCPIs for the first year or so.

The old Citroen gearbox had served Lotus well and was still available from France, but all the mainstream Citroen/Peugeot models which had used it had dropped out of production. Soon, Lotus reasoned, the French company might want to close down supplies.

The 25/GTA-type gearbox, on the other hand, was relatively modern, and had an assured future. This gearbox, incidentally, was a very versatile unit. In the sporting GTA itself it was arranged to be *ahead* of the final drive unit, and driving back to it. In the Renault 25, it was *behind* the line of the final drive, and driving forward to it! For the new Esprit, therefore, where the gearbox was behind the final drive, there were strong similarities to the Renault 25 type of installation.

Here, for comparison, are the ratios of the new transmission and the old:

	Renault-type gearbox for 1988 Esprit	Citroen-type gearbox for 1975–7 Esprit
Internal ratios:		
5th	0.76:1	0.82:1
4th	0.97:1	1.03:1
3rd	1.32:1	1.38:1
2nd	1.94:1	2.05:1
1st	2.92:1	3.36:1
Reverse	3.15:1	3.54:1
Final drive ratio	4.375:1	3.889:1

The internal gearbox ratios in the Renault 25 *and* the Renault GTA incidentally, were exactly the same as those supplied to Lotus, which made parts supply even easier to ensure. The V6-engined Renault 25 also had the same final drive ratio, but that of the Renault GTA was significantly higher.

There was one other significant mechanical change, which was forced upon Lotus. When Renault were developing the new gearbox, they had decided to use outboard disc brakes on their cars. Accordingly, there was no way that Lotus' inboard discs could be fixed to the new casing. The 'old' Esprit, therefore, had always used inboard disc brakes, fixed to a Citroen transmission intended for such a fitting; the new car had outboard discs for the very first time.

Clearly the Esprit restyle had arrived at exactly the right

The X180 Esprit had a new type of instrument panel and facia, though the general layout and features were familiar. The footwell package was more spacious than before – but all such improvements were relative, as it still helped to have small feet and relatively short legs.

Minor body changes to the Esprit S2 included a more integrated front spoiler and neat air intakes into the engine compartment positioned at the trailing edge of the side windows. (Picture: Focalpoint)

A limited-edition version of the Esprit S2 in JPS black-and-gold colours was introduced to commemmorate Lotus' victory in the 1989 Formula 1 World Championship with Mario Andretti, to whom this particular car was supplied. (Picture: Focalpoint)

moment. Backed by the might of General Motors, Lotus' self-confidence was glowing as never before, and they had recently acquired control of the Lotus sales/distribution organization in the USA. This was moved from New Jersey (west of New York) to a brand-new facility at Atlanta, Georgia.

The company was making good profits, the still-secret new Elan was on the way – and sales shot up. At the London Motorfair in October 1987 Lotus announced that they had taken orders for 371 Esprit Turbos, worth £8.2 million.

In 1987 a total of 462 Esprits – new type and old type combined – had been produced, but helped along by the new style this total rose to no fewer than 1,058 in 1988, which was easily an Esprit record. In fact this was an all-time Esprit record, for although there would be another major new development in 1989, Lotus never again built more than 1,000 Esprits in a year.

In the first half of 1988 UK sales totalled 271 cars, while exports rose to 311, of which no fewer than 172 went to the USA. At the end of 1988 those figures had risen to 572 UK sales and 323 USA sales.

The new-style Turbo was a genuine 150mph car and this, allied to overall fuel consumption of around 20mpg (Imperial), made the latest car an intriguing proposition. Purists still complained about the restricted cockpit space, the need for power-assisted steering and the fact that the car did not have ABS anti-lock braking, but even they could drive the Esprit faster than almost every other supercar, and still arrive at the end of a trip with a big smile on their faces ...

As *Autocar* testers wrote in April 1988: 'The Esprit Turbo has many points in its favour. Searing performance, a chassis and brakes to match and sensational looks are only part of it. You also get acceptable fuel economy, a fine ride with very little road noise and reasonable luggage space for a two-seater ... By supercar standards, the Esprit Turbo is something of a bargain.'

A month later *Road & Track* tested a USA-specification Turbo, headlining its report: 'Hethel's hardy perennial blooms again'. There was a lot more: 'Throw the Lotus into a corner and the first thing you notice is – nothing. Other than quick, accurate steering ... It's an exotic toy for people who not only can afford it, but also will appreciate its particular brand of performance as an extension of the Lotus tradition, and a handsome one at that.'

By this time the Esprit Turbo had taken over the majority of sales in all markets. It was, of course, the only Esprit to be sold in the commercially important North American market, yet it was now such a success that it was overshadowing the normally-aspirated Esprit almost everywhere. Sales of normally-aspirated cars gradually fell away – from 176 in 1988 to 90 in 1989, and to a mere 22 in 1990, after which the model was allowed to fade out gracefully.

The development engineers, on the other hand, redoubled their efforts to make the turbocharged cars ever better. In 1988, 1989 and 1990 there would be major improvements to the engine and to the chassis – and there was even time for a special Italian-market model to be launched, too.

It was remarkable, at a time when the front-wheel-drive Elan project was being urged towards its launch, that there was still opportunity to work so hard at an established design. The important chassis change came in mid-1989, when what Lotus colloquially call the 'Eagle' chassis was introduced. The basic design was not changed, but there were revisions to suspension geometry which eliminated anti-dive, and a reduction of the front castor angle. Along with an increase in bump travel, the use of stiffer front springs and 16in diameter rear wheels – but the retention of 15in front wheels – this showed signs of a radical rethink.

Turbocharged engine developments

The work which went into engine improvements in the same period needs explanation. According to the list supplied by Lotus, 10 different versions of the turbocharged car were produced between 1987 and 1991, along with the much-modified X180R SCCA race car, which is covered more fully in Chapter 7.

All are described in Appendix A, but to relate one to another, here is the sequence of events. 'Federal' markets mean the USA and other countries which impose similarly strict engine emission standards. From 1990 the 'Turbo' word was abandoned, but even on sober reflection the badging policy was confusing:

October 1987   Launch of two X180 types:

Turbo HCPI for 'federal' markets. With 215bhp, Bosch injection and old-type Citroen gearbox Turbo for other markets, also with 215bhp, Dellorto carbs, more torque, and Renault gearbox.

Late 1988   Turbo (MPFI) replaced Turbo (HCPI) model. Revised engine, with multi-point Lotus/Delco fuel injection and 228bhp. For sale in federal markets. Fitted with Renault gearbox (Citroen gearbox no longer used on any Esprit).

Mid-1989   Turbo SE launched as additional model. With charge-cooled engine, Lotus/Delco engine and 264bhp. New 'Eagle' chassis introduced to all Esprits at this point.

Autumn 1990   Esprit (normally-aspirated) discontinued. Esprit Turbo renamed Esprit (confusing!), with 215bhp engine, but with open-back tailgate and other style 'cues' from obsolete normally-aspirated model. UK market only.
Esprit S introduced with 228bhp engine, MPFI and glass-back style. UK market only. Available until 1991 only.
Esprit (turbocharged) introduced with 228bhp, MPFI and glass-back style. All export markets. Available until 1991 only.
Esprit SE was new name for Esprit Turbo SE, for sale in the USA and all right-hand-drive markets. 264bhp.
Esprit name was given to 264bhp charge-cooled car for left-hand-drive markets.
'SE' designation dropped to avoid trademark infringements.

ABS anti-lock braking was standardized on all models.

1991   Esprit 2-litre launched, for sale initially in Italy. With 1,973cc/250bhp, engined otherwise as for Esprit SE.

In the period just detailed, the first major change was to specify a different type of fuel injection system – what was called Lotus/Delco MPFI (Multi-Point Fuel Injection) instead of the original Bosch system. The key to this change was the word 'Delco', for Delco was a familiar General Motors brand name, and Delco systems had been in use on North American GM engines for some time.

Even more than the Bosch K-Jetronic system, which was a long-established fuel injection package used by many other car makers, this was attuned to all modern and demanding exhaust emission requirements. When the time came, too, GM and Delco would be able to cope easily with the requirements for catalytic converters.

It was a great bonus for Lotus that the Delco system also liberated more power (228bhp vs 215bhp) *and* more peak torque (218lb.ft at 4,000rpm compared with 192lb.ft at 5,000rpm) than the Bosch system had done.

The major leap forward, however, followed in mid-1989, when Lotus finally made a charge-cooled engine available. To be frank, it was long overdue, for although the company had been one of the very first to put a turbocharged road car on sale, they had then fallen well behind other manufacturers as they refined the technology further.

The theory of charge-cooling, or intercooling, is simple enough. Compressed air forced out of a turbo compressor is hot, and if it can be cooled before entering the engine the volumetric efficiency is dramatically improved. Cooling is done through a radiator – either an air-air type, or an air-water type.

As might be expected, the Lotus installation was neat and effective. In this case the charge cooler was mounted on top of the engine, and the coolant was water, this being in a self-contained system, circulated by an engine-driven pump to a front-mounted radiator.

This, however, was merely part of a comprehensive improvement package, which also included Lotus/Delco injection and the ability to run on lead-free fuel, a catalyst being

The introduction of a longer-stroke version of the Lotus 2-litre engine in 1978 led to the new Esprit model designation S2.2. Galvanized chassis treatment was introduced at the same time. (Picture: Focalpoint)

The garish colours of this 1980 limited-edition Esprit Turbo provided a reminder that Essex Petroleum had replaced John Player as sponsors of the Lotus Formula 1 team. (Picture: Focalpoint)

From this angle the restyled X180-type Esprit looked considerably more plump than the previous type. The glass roof panel and the free-standing rear aerofoil (available from late 1988, standard on the Commemorative Edition) are both obvious. And if you're impressed by a new style of rear-view door mirrors ...

standardized. Not only did all this help push up peak power to 264bhp (DIN) at 6,500rpm, but there was also a transient over-boost facility which allowed usage of 280bhp at 6,500rpm for up to 30 seconds of hard acceleration.

Two years after this charge-cooled engine had been launched, a further variant was introduced, this being a 2-litre version, initially for sale only in the Italian market. The rationale was fiscal, not technical, for in Italy much higher motoring taxes were applied to cars with engines of more than 2,000cc.

There were precedents for this move – Ferrari, Maserati and Lamborghini had all offered smaller-engined versions of their cars from time to time, though in no case had sales come up to expectations. The development cost of producing a smaller-capacity engine – which actually used the self-same crankshaft as the original Type 907 engine had done in the Seventies – was small.

With the introduction of the charge-cooled turbo engine,

Lotus claimed the title of the world's highest specific output for a road-car engine (121bhp/litre), though this was not likely to be held for long.

By the early Nineties the famous 16-valve engine had been on the market for 20 years, but Lotus clearly thought more potential was still locked in there. In 1992, as this book was being prepared, the first of an entirely new design of aluminium cylinder block, more rigid than ever before, was phased into production.

Esprit improvements – 1988–92
In the spring of 1989, when the charge-cooled engine was put on sale, Lotus also introduced a series of chassis and engine changes to the Esprit. The 'Eagle' chassis, so called because of the exclusive use of Goodyear Eagle tyres, which is easily identified by its use of 15in diameter front and 16in rear wheels, was standardized on all Esprits at this time, but the derivative with the 264bhp charged-cooled engine became

the Turbo SE (SE = Special Equipment). The cost was a massive £42,500.

The SE not only had far more performance – a top speed of 159mph and 0–60mph in 4.9sec, which was remarkable by any standard and equal to the performance of the Porsche 911 Turbo of the period, but there had been further changes to the aerodynamics. At the front there was a deeper front 'bib' spoiler, the sills and their associated air intakes to the engine bay had been reprofiled and, for the first time on an Esprit, there was a small free-standing spoiler on the tail. Air conditioning and a tilt/removable glass sunroof were also standard.

Inside the cockpit, there was more obvious luxury – perhaps an attempt to justify some of the extra cost of the car. As *Autocar & Motor* stated in its road test: 'The SE's all-leather cabin has more of a "quality" ambience than the regular Turbo's, an impression helped along by the large slab of burr elm surrounding the instruments and switches ...' This, by the way, was another enthusiastic road test: 'With the Turbo SE, Lotus has made a break for the big time and has succeeded admirably in all the most obvious respects. To many eyes the Esprit has always looked the part and now those looks are matched – even surpassed – indeed. The SE is fabulously rapid and enormously capable ...'

The next batch of important changes came in the autumn of 1990, effectively for the start-up of the 1991 model year. This was the point at which the normally-aspirated Esprit finally died, and when a whole series of model name changes confused the historians, particularly if home market and export market titles were compared ...!

The most important mechanical improvement was that three-channel ABS anti-lock braking was finally adopted – though not the power-assisted steering for which the pundits had been nagging for some time – while for the British market this signalled the point at which three different models were briefly available. Note the model names – carefully:

Esprit	215bhp, Dellorto carbs	£34,900
Esprit S	228bhp, Lotus/Delco injection	£38,900
Esprit SE	264bhp, charge-cooled, Lotus/ Delco injection	£46,300

By this time, however, the Lotus factory and the Lotus dealers could no longer give the Esprit their full attention. They were too bound up in the front-wheel-drive Elan, which had been on volume sale since April 1990, to think of much else. This was also the point at which British economic recession began to bite in earnest, and when Lotus *should* have realized that the Esprit was beginning to look expensive. Perhaps this also explains why the Esprit S had such a short life – for it was dropped after only a year.

Twelve months later, and at a point when the front-wheel-drive Elan still appeared to be selling very well indeed, more Esprit changes were phased in. Not only was there a very significant improvement to the cabin packaging of all types, but the SE model received an extrovert new styling package which was claimed to reduce understeer and raise the top speed.

From this point, all Lotus cars were provided with a three-year unlimited-mileage warranty – experienced, not to say cynical, Lotus owners thought this was likely to be a *very* expensive deal for Lotus! – and no doubt partly to allow for this the UK prices were raised yet again, with the price of the flagship SE model soaring to £48,260.

Somehow or other the cabin had been repackaged and made more spacious. Compared with earlier cars there was an extra 5.6in of headroom, 1.6in more legroom and up to 3.2in more clearance between the seat cushion and the steering wheel rim. A repositioned cabin/engine bulkhead made the cabin 1.2in longer overall. There was a revised pedal box which gave shorter pedal movements and the centre tunnel had been slimmed down to allow wider seats to be fitted.

Added to this was the provision of doors which opened by an extra 15deg – and if this does not sound much, it equated to an extra 9in at the rear end of the doors.

The principal visual changes were only applied to the SE type, the official reasons being twofold: to add front-end downforce to cut the understeer, and to improve rearward visibility. The glass-back which had featured on X180 types since 1987 was abandoned, the original small spoiler of the SE was no longer fitted, and in their place was a larger and free-standing rear aerofoil, mounted on pylons. To balance

The Esprit received substantially changed bodywork in 1988, offering a more rounded profile as well as improved aerodynamic drag. Several variations of rear wing for both the SE and Turbo SE were to be seen over the next few years. (Picture: Focalpoint)

The Esprit SE Turbo in 1992 guise with its dramatically high-mounted wing supported by short struts over the tail of the body and long arms extending from the back of the cabin. (Picture: Focalpoint)

The last of the normally-aspirated Esprits were built in 1991, bringing to an end the 16-year career of this particular type of Lotus chassis.

this, at the front there was an extra rubber lip under the existing 'bib' spoiler.

Lotus claimed that the overall effect was to improve downforce at higher speeds, to cut the drag and to raise the top speed to no less than 165mph.

Even so, by mid-1992, when the Elan was suddenly killed off and the last Excels had also been built, the Esprit had faded quietly into a limited-production backwater all of its own. Yet again sales in the USA had collapsed, and the British recession had hit hard at sales of all high-performance cars, resulting in Esprit production being cut back to no more than five cars a week.

Except for the Lotus-Omega/Lotus-Carlton assembly shop, by the autumn of that year the Esprit was the *only* Lotus car in production at Hethel, but in such low numbers that the factory seemed almost to be a ghost town.

What future for the Esprit?

In a book like this, and especially where Lotus are concerned, making predictions is a dangerous business. Accordingly, I can only summarize where the Esprit stood as the end of 1992 approached.

Immediately after the Elan had been killed off in June 1992, Lotus' Managing Director, Adrian Palmer, made it clear that Esprit assembly would continue indefinitely, and that several significant developments were already in the pipeline.

On the one hand, the rumour-mongers told us, even faster 'super-Esprits', with different and more powerful engines, would be announced. That, however, was not likely to increase the volume of sales.

The other strong possibility, it seemed, was for an 'entry-level' Esprit to be developed, one that had a much cheaper engine, probably the first not to have been built at Hethel. If the car's retail price could be reduced significantly, this might boost sales quite dramatically.

Favourite, it seemed, was the brand-new 2.5-litre 54deg V6 unit which General Motors had just revealed, and which would go into series production in 1993. This was originally intended to power Opels and Vauxhalls, and was a torquey, normally-aspirated engine with 167bhp.

In the early Nineties it seemed as if Lotus were determined to maintain the Esprit pedigree, and to improve it further in future years.

By 1992 the Esprit Turbo SE had been further modified, with a more spacious interior. It had also been given a large and aerodynamically-efficient rear aerofoil to improve its stability. But it was a short-lived model, for early in 1993 this S4 version was unveiled, with an even smoother profile and important chassis changes which were to transform the Esprit's handling at a cost of a somewhat firmer ride.

Esprit for the 90s. The S4, which replaced the S3 in 1993, can be identified by its different rear wing, nose and five-spoke wheels. It was also given important suspension revisions which transformed the handling. (Picture: Focalpoint)

CHAPTER 6

Esprits in motor racing

From SCCA to IMSA

Lotus had never been involved in race-tuning the Esprit. Team Lotus was wholly occupied with the Formula 1 scene, and there was no other works team as such at Hethel. From time to time private owners prepared Esprits for production and modified-production sportscar racing, but the factory did not back them – until the late Eighties. By then General Motors had taken control of Lotus, the new M100 Elan was on the way and the North American sales organization was pushing hard to raise its profile.

The problem at that time was that in North America Lotus still had a serious image problem. The only Lotus model currently on sale was the new X180-style Esprit Turbo, a car which the buying public saw as very beautifully styled, very fast and with excellent roadholding – which was the good news. The bad news was that there were still questions about durability, reliability and parts supply.

To kill that image, Lotus Cars USA reacted in the predictable – and aggressive – way by getting into motor racing with the Esprit. All 70 USA dealers were asked to support a racing programme instead of a massive advertising campaign, and the majority agreed that a successful return to Lotus' racing roots was an ideal way of generating a good image.

Lotus soon decided to hire Pure Sports Inc to prepare a small team of works race cars to be entered in the Sports Car Club of America (SCCA) Escort World Challenge. In many ways the SCCA is the equivalent of Britain's BRSCC: ubiquitous, well-organized and with race meetings promoted all over the very large continent. Success by the Esprits would reach a very large and influential audience.

This was not a money-no-object series for 'homologation specials'. If it had been, American manufacturers would surely have lost out, for the black art of homologation was something practised *outside* the USA. Like many other types of North American motorsport, this series was set up as a showcase for manufacturers who were selling cars in North America, the rules insisting on the use of standard components which could be strengthened or otherwise upgraded.

In an astonishingly short time, ex-F1 driver – and Lotus chassis consultant – John Miles led a team at Pure Sports which made two cars ready in time for the first race in March 1990. Very few shakedown laps and virtually no development testing was possible at the Smithton circuit. Merely five weeks after approval for the programme had been given, 'Doc' Bundy and Scott Lagasse shared a car in a three-hour SCCA race at Sears Point – and won outright! Contrary to all the cynics' expectations, nothing fell off the car – and the Esprit defeated every Corvette, a model which had dominated SCCA racing for several years.

This was an astonishing debut for an unproven car in an unproven team. Competition, from factory or importer-sponsored teams of Chevrolet Corvettes, Nissan 300ZX Turbos, Porsches, Mazdas, Hondas and BMWs was intense – but the Corvettes were the fiercest competitors.

In 1990, the first year in which Lotus were ready to take part, races varied in length from 45 minutes to three-hour endurance events – and there was one incredibly gruelling

The Lotus Sport Esprit Turbo, as raced in North American IMSA events in 1992, when Doc Bundy won the Driver's Championship. Although it was an extensively race-modified car, the same basic chassis, body and running-gear as all road-going Esprit Turbo SEs was used.

24-hour race at Mosport Park, in Canada. It was a season which had its ups and downs for the Esprits (the second car appeared in time for the second race). The new cars won four out of eight races entered, the statistics including six pole positions, six fastest laps and 2,900 racing miles, but one car was written off in an accident with one of the Corvettes, there were problems with the ABS system – once fitted – and throughout the season the main hassle came from the gearbox and the wheel bearings.

Even so, Doc Bundy finished second in the Driver's Championship, and Pure Sports was awarded the Jim Cook Memorial Trophy for its 'significant contribution to the overall success of the Championship series, and consistent display of good character and sportsmanship'.

In 1990, these were the principal changes applied to the Esprit Turbo SE specification to turn the model into an incredibly effective race car:

Engine: Blueprinting, power not quoted.
Chassis: Ungalvanized, with sturdy rollcage welded into place.
Suspension: Revised front geometry, adjustable camber at front and rear, uprated suspension bushes, revised spring, damper and anti-rollbar rates, with a rear anti-rollbar fitted. Ride heights lowered by 10mm.
Brakes: Lotus-Omega/Carlton type used (13in front discs, 12in rears), with 4-piston alloy calipers. Handbrake deleted. ABS fitted in mid-season. Cooling trunking fitted at front and rear.
Wheels/tyres: Goodyear Eagle ZR-rating used all round on 16in rims. 225/50-section fronts on 8.5in rims, 225/45-section rears on 9.5in rims.

To meet SCCA Championship rules, changes were made towards the end of the season, the Esprits having to weigh 2,600lb – 100lb higher than at first – which caused tyre wear and tyre carcase problems, especially in the long races.

Even before the end of the first year, Lotus had reacted to their racing success by deciding to build a series of 20 'street-legal' racecar replicas. These machines, priced at a massive $125,000, were ready for Lotus-USA dealers to sell in the winter of 1990–1, and were known by factory personnel as the X180R (R = Racing ...) models.

Although these cars (specification listed in Appendix A)

At the NEC motor show of October 1992, Lotus showed a development prototype of the Esprit Turbo called the Sport 300. This was a lightweight ultra-high-performance derivative of the X180R racer, with 300bhp, a limited-slip differential, wide-section tyres and 17in diameter rear wheels. No price was quoted at that time, but Lotus claimed that it was already available to run, and was road-legal in most European countries.

had standard 264bhp engines, galvanized chassis, standard bodyshells, removable roof panels and heaters, they also had rollcages, racing wheels and tyres, uprated suspension yet standard bushes, special sports seats, but a revised bib spoiler which incorporated front wheelarch air deflectors. As such they were no faster than standard, but were visually stunning.

1991 – Pure Sports out, Lotus Sport in

For 1991 Lotus decided to make a full-scale assault on the SCCA title *and* to tackle the IMSA Supercar series, which had hitherto been dominated by works-backed Porsches. To this, a new team called Lotus Sport was set up in Atlanta, which built three new cars and hired four new drivers, one of whom was film megastar Paul Newman, another Doc Bundy, their partners being Bobby Carradine and Michael Brockman. Although Newman was a real sportsman and a great person to have around, he was no longer a young man and was not as quick as the other drivers. Carradine never finished a race out of the top five positions, and had one outright victory, while Doc Bundy won twice.

Compared with the 1990 cars, the new machines had a more complex and rigid rollcage, increased suspension geometry adjustment all round and even larger wheels and tyres. Fronts were 245/46-16in on 8.5in rims, rears 315/40-17in on 10in rims. There was an X180R bib spoiler, and the engines produced 300bhp.

For the SCCA series the Esprits retained Goodyear Eagle tyres, but the IMSA series specified the use of Bridgestone tyres by all competitors, which meant that there were rather different set-ups for the two types of racing.

In 1991 the Esprits took part in nine races. One particular weekend saw a SCCA race on the Saturday and an IMSA race on the Sunday, both at Elkhart Lake, Wisconsin! In a concentrated season the car won four races, finished second three times and third four times; only once did the team fail to finish in the first three. As in 1990 the team finished second in the SCCA series, but in 1991, to their utter joy, they also beat the Porsches in an IMSA race on their home territory at Road Atlanta. Once again the Jim Cook trophy was awarded to Lotus.

Even though Lotus' North American operation was still in serious financial trouble – for by this time the M100 Elan's sales failure was only too evident – the company mounted another serious attempt in 1992. Because John Miles had joined the Team Lotus F1 effort, Roger Becker became the director of this programme.

Lotus Sport's third season concentrated on the IMSA Bridgestone Supercar Championship, an eight-race series for production sportscars which received national TV coverage. The 1992 Esprit Turbo X180R was a further refined version of the 1991 cars, this time running with Bridgestone Potenza tyres, which were the 'control' tyres for the series. The engine had been further developed to give 335bhp, and because of more favourable regulations the weight had been trimmed to 2,400lb.

Yet again there had been a change of wheel/tyre specification, confirming just how versatile the Esprit's chassis was, even though it was by no means a new design. For 1992 there were 225/60-16in front tyres running on 8.5in rims and 275/40-17in rears running on 9.5in rims – which left a stock of surplus 10in rims back in Atlanta!

In a busy season which saw four cars used, and with Paul Newman still behind the wheel of one car at the third time of trying, Doc Bundy went on to become the Bridgestone Supercar Driver's Champion, clearly defeating Hurley Haywood, who had been driving a works Porsche.

This was an excellent performance, especially as the Bridgestone series included works-supported cars from Chevrolet (Corvettes), Porsche and Mazda. The race for the title soon developed into a two-man contest between Bundy and Haywood, which was only settled at the final event near San Diego in California, in October 1992.

At the end of that triumphant season, Roger Becker said: 'We've a heritage of track success. These victories show how potent and reliable today's Lotus Esprit is, both for track and road use. We plan to expand our competition programme next year.'

At the end of a year which had been singularly traumatic for Lotus, those were indeed brave words. If the Esprit was forced out of motor racing it would be because of a lack of money, not a lack of success. If only the Esprit Turbo had started racing several years earlier. If only ...

CHAPTER 7

Etna and V8s

Prototypes and contract work

During the Eighties Lotus not only built a number of exciting prototypes, but became increasingly involved in engineering contract work for clients all over the world. On the one hand there was work on cars such as the Lotus 'Etna' project, which *was* made public, while for other clients, suspension, engine and structural work was carried out. After the Elan project had been stripped out of Lotus in 1992, the engineering side of the business became the dominant part of Group Lotus' operations.

This chapter, therefore, surveys some of the work – that which was made public – on a few of the projects which occupied Lotus during this time.

Etna, M300 and V8 engines

As long ago as the late Sixties, Lotus conceived a new series of engines – the 900 family – which encompassed a four-cylinder in-line unit and a closely related 90deg V8 unit. The fact that the four-cylinder engine was always intended to be installed with its cylinder axes leaning over at 45deg made it clear that a 90deg V8 could eventually follow.

To hope to design such an engine, however, was much easier than to build, tool for and fit it into a production car. Even by 1970, type numbers for a 32-valve four-cam V8 engine had been allocated – Type 908 referring to a racing 4-litre and Type 909 to a road car derivative – but actual prototype engines were not built until the end of the decade.

During that period Lotus directors talked wistfully of the desire to fit a V8 into a production car – the engine, as packaged, could certainly have been fitted to any of the Elite/Eclat/Excel/Esprit generation of cars – but little concrete progress was made.

In many ways the design of the V8 engine was simple enough to achieve, for it meant laying out a new aluminium cylinder block to accept two of the proven 16-valve twin-cam cylinder heads. The bore and stroke – quoted as 95.28mm x 70.3mm – were similar to those used in the four-cylinder engines – the stroke was slightly longer – which meant that the cubic capacity was 4,010cc. With a compression ratio of 11.2, doubling up the 160bhp of the 'four' gave a nominal power output of 320bhp, though more was certainly available if required. It is now known that it would have been possible to enlarge that capacity to 4,348cc by using the 2.2-litre four-cylinder engine's dimensions, and there might have been just a little more to come.

As Tony Rudd later made clear, actual engine design began in 1979, and by 1982 six prototype engines had been built and tested with a variety of induction systems, but at that stage there was no use for the units. Lotus could not afford to design a new mid-engined car to use the V8 engine, and no sales to contract customers were on the horizon.

Finally, in October 1984, the Etna project car was unveiled at the Birmingham (NEC) motor show, though this was still more 'project' than 'car'. A new V8 mid-engined Lotus production car would have been coded M300, but little had been done about a settled design for such a car. At this time, in fact, there were several paper projects floating around at Lotus, for M300 was also joined in the minds of the designers by M80 (or 'Eminence'), being Lotus' idea of a luxurious V8-

Years of yearning finally came to fruition in 1984 when Lotus showed the mid-engined 'Etna' concept car, the first and only Lotus to be powered by a Type 909 V8 derivative of the famous four-cylinder 16-valve Esprit engine. Only the one car was built.

powered four-door saloon to rival Mercedes-Benz and Jaguar.

Although the layout of this car was basically similar to that of the Esprit, it had a 3in longer wheelbase (99in instead of 96in), was 14ft long, powered by the new 4-litre Type 909 V8 engine and had a startling but somehow quite bland two-seater coupe style by Giugiaro. Naturally, this was wedge-shaped, with a very low front end.

In 1984 Lotus saw Etna as a 'realistic and practical prototype of the future', and its specification included not only the 4-litre V8 engine and bodywork intended to be made by the VARI process, but it also had provision for a new type of CVT automatic transmission, Active Ride suspension, power-assisted steering and ABS anti-lock brakes. Some observers criticized the style as being too predictable, a cynic once commenting: 'Ah yes, pure production-line Giugiaro, he has a cupboard-full of shapes waiting for a client to ask about ...', but it looked good and had a claimed Cd of only

0.29. With the quoted 320bhp, Lotus had already computed a top speed of 182mph and 0–60mph acceleration in 4.3sec.

The interior was a development of that of the Esprit, with a cascade of controls and switches on the centre console which were sharply angled towards the driver. This console, therefore, was heavily handed, which meant that different mouldings would be required for right-hand and left-hand-drive cars.

The Etna shown in 1984 was incomplete – I do not think it ever *was* finished or driven – although a V8-engined Esprit prototype with a five-speed manual transmission had already been built and run. The press was never shown the engine bay. At the same time, however, Lotus released one picture of the engine in which there were eight inlet port stubs, but no sign of carburation or fuel injection systems, though three different types of fuel injection and a carburettor system had already been tried. It was claimed to weigh just 414lb, with Lotus making the astonishing, unbelievable claim that it was actually 4lb lighter than the Type 907 four-cylinder engine!

At the motor show, Chairman David Wickins reaffirmed that Etna could become a production car by 1988, but that its future depended on the provision of funds which would still have to be secured from new backers: 'We would not have to produce many to make money. All the development costs on both the car and the V8 engine have already been written off. So if we stopped now, we would lose money ...'

There was also apparently the possibility of the engine being sold for use in a non-Lotus car: 'One manufacturer in the USA has already said "No" because the engine is *too* powerful. He reckoned it would just break his cars apart!'

Unhappily, little more was ever heard about Etna, and nothing was ever officially revealed about M300. In 1985,

when Peter Stevens began work in earnest on a new small car, the X100, which eventually became the M100 Elan, he was influenced by the shape of the Etna but, according to Elan archivist Mark Hughes: 'With Etna looking less likely than ever to reach production as the months went by, the styling links began to seem pointless.'

By 1986, in fact, Lotus' efforts were increasingly channelled into developing the new Elan project, which resulted in Etna being cancelled, and the V8 engine with it. M300, though a bold look into Lotus' 'Supercar future', also never came to anything, for even in 1989, when the company had just announced the M100 Elan and were at their most bullishly optimistic, the M300 was forecast as being a 1993 project and never got beyond the 'who don't we ...?' stage.

The V8-engined Etna's facia/instrument panel was a futuristic development of what Lotus were already producing in the Esprit. But how many of the instruments actually worked, and how many were functional?

This is the only picture of the Type 909 V8 engine which was ever published – in 1984, when Etna was also unveiled. There was no sign of any carburation or fuel injection. Running prototypes used several different types, though none has ever been seen in public. The likelihood of this engine ever being put on sale is remote.

When I was allowed a look behind closed doors at Lotus in 1992, I saw several examples of the V8 engine in store, with differing fuel injection installations. Most intriguingly, I also saw a single example of a V12 engine which was quite obviously a direct derivative of the V8! The visual contrast with a tiny V6 turbocharged 1,000bhp F1 engine for Toyota, stored alongside, was startling!

This V12, clearly, was a 6-litre engine, which must surely have developed between 450–500bhp, and would have outranked almost any other V12 engine in the world if put into limited production. The story is that Lotus produced it at General Motors' behest for fitting to a front-engined 'Super-Cadillac', but that this project was cancelled at an early stage. The engine was then somehow shoehorned into an M300 mock-up, in a transverse position – the car would

apparently have been much wider than a Ferrari Testarossa – but according to stylists who never worked on it, it looked 'appalling'.

No-one at Lotus was willing to say any more, but remember – you read it here first ...

Active Ride suspension

Early in the Eighties Lotus set out on a programme of developing what is known as Active Ride suspension – a system where complex hydraulic and electronic controls can sense the tendency of a vehicle to roll or pitch as soon as it begins, and react accordingly by resisting roll in the correct way. The elusive reward – to use a well-worn cliche, the crock of gold at the end of the rainbow – is to develop a car which does not roll or move around, but which still provides a comfortable and stable ride for the occupants. In practical terms, a car could also be made more compact, for with minimal wheel movements required, less space would be needed for large wheelarches and suspension linkages.

Like almost every other car manufacturer, Lotus found it easier to discuss the technology and to build one-off prototypes than to evolve a cost-effective system which would work all the time. Ten years after work began on Active Ride, it seemed to be little nearer production, and except for research work for outside clients it no longer seemed likely to find a place in a Lotus car. Certainly, it was unlikely that Active Ride would be fitted to any surviving member of the Esprit family in the Nineties.

Lotus first developed Active Ride for their F1 single-seaters to allow these cars to get the best out of their tyres and to optimize the ground-effect properties of their cars. On the F1 cars, as on later road cars, the idea was to use Active Ride to keep the structure of the cars in a constant relationship to the ground.

Progress in F1 can best be described as spasmodic. The first 'Active' cars, the Cosworth DFY-powered Lotus 92s, were raced in 1983 with a conspicuous lack of success, and although Lotus persisted in threatening to standardize a developed system on later cars, this never actually came to anything. Even in the early Nineties, when Lotus F1 strategy was in the hands of Peter Collins, and with engineer John

Miles back with the team, the latest Type 107s were still what is known as 'passive' cars.

In 1983, however, Lotus made much of the prototype Active Ride systems being developed for and fitted up to the Excels and Esprits. The very first Esprit ran in March 1982, and the public unveiling came in September 1983.

The basis of the system was four Eland hydraulic jacks, which replaced the conventional road springs and dampers of the car. In a further developed system they would also eliminate the need for anti-rollbars. Oil could be admitted at either side of the pistons, via electrically operated servo valves – made by aerospace specialists Dowty – which could cycle incredibly quickly – 250 times *per second*; conventional car dampers rarely cycle at more than 20 cycles per second.

So far, it was so simple, but the complex, high-tech, super-advanced – call it what you like – part of the Active Ride was the control of those oil movements, and the sensors which activated them. The first Esprit had 14 main sensing devices, which detected ride heights, wheel loadings and the lateral and longitudinal acceleration of the vehicle.

The aim was to produce a vehicle which would not roll on corners, dive under braking or pitch up under hard acceleration. Each and every one of these was to be controlled automatically, the passengers not to be made queasy by it all, or even to notice what was happening. As development advanced it was intended to build in the degree of understeer and other subjective feelings which were right for the car and its character.

The aim was laudable, but as the years passed it proved to be inordinately difficult to provide at any reasonable cost. As John Miles, then a technical writer with *Autocar*, commented in 1983: 'Whether the electronic and hydraulic systems could ever be developed down to a price that would make active suspension of this type a mass production possibility is open to question, especially if aircraft-like self-checking and back-up systems had to be included on safety grounds ...'

Years later, when he went to work for Lotus, he discovered just how wise his earlier words had been!

In happier and more expansive times, Lotus might have produced a car like this. Eminence was a design study for a Type 909 V8-engined four-door five-seater supersaloon car – but it never progressed any further than this 1984 drawing.

Microlight projects

For a time in the early Eighties Colin Chapman was keen to see Lotus diversify into the production of lightweight aircraft engines and the airframes to go with them, and the very first of these, having been revealed during the summer of 1983 – when the two-seater plane was tested by Dick Rutan in the USA – flew just one week after Lotus' founder died.

In October a flat-twin air-cooled engine was exhibited at the British Motorfair at Earls Court, when it was suggested that the company had orders for 2,500 units. This was a tiny four-stroke 480cc engine which developed 25bhp at 5,000rpm, and before long it was joined by a 50bhp 960cc flat-four, which was effectively a doubling-up of the original design.

The twin was apparently scheduled for use in the Eipper-Lotus microlight craft, a machine designed by Dick Rutan and Scaled Composites Inc and featuring a strikingly-styled canard layout, with the tail fins actually at the nose of the fuselage.

Here such a project leaves the scope of this book, and I believe that little more was ever heard of it in the world of light aviation.

LT5 – a magnificent V8 engine for Chevrolet

Over the years Lotus engineers have designed any number of amazingly powerful engines. There are individuals who boast that Lotus are as resourceful as Cosworth, but that Lotus never gets the credit for what they have done.

Some engines, like the Formula 1 1.5-litre V6 turbo unit for Toyota, were never revealed; some, like the 3.6-litre twin-cam for the Lotus-Carlton, were officially adopted by General Motors; and others never made it beyond the drawing stage. The LT5, however, was a copybook design from beginning to end. Started in 1975 and revealed in 1989, it made the Chevrolet Corvette ZR-1 the fastest production

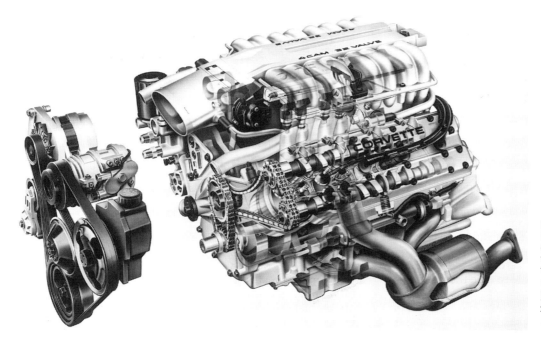

Even before being taken over by General Motors, Lotus began designing a magnificent new 32-valve, four-cam V8 engine for use in the Chevrolet Corvette. This, the LT5 unit, was a 375bhp 5.7-litre master-piece, and powered the Corvette ZR-1 which went on sale in 1989. The top speed was more than 170mph.

car ever put on sale by General Motors.

As may be expected of an engine designed to fit into an American car, LT5 was a V8. What may not be realized is that it was really a technical update to the thinking behind the Lotus Type 909 V8 of the early Eighties. There were no common components, but the basic layouts were similar, and one was a definite descendant of the other.

For Technical Director Tony Rudd the story began in 1983, after Colin Chapman died: 'We had six V8 engines and no money. I tried to sell the design to anyone who'd listen, and I spoke to Russ Gee, head of powertrains at Chevrolet/Pontiac, Canada. He was politely interested, but GM was being restructured.'

Then GM instructed Gee to make a European tour, to search out a new engine design – one which could power a 'super-Corvette', and could even be raced in the USA. Against stiff competition from Cosworth, Ricardo and Porsche, in April 1985 Lotus was commissioned to design new four-valve-per-cylinder twin-cam heads on GM's existing L98 'small-block' 5.7-litre V8 engine, which was a classic 'Detroit-iron' unit. This was nine months *before* GM took over Lotus, that commercial move coming in January 1986.

Initial work on this conversion – for that was what it really was – revealed an installation problem. Corvette assembly involved the engine being offered up from below the bodyshell unit – and this could not be changed – which put a limit on the overall width of the twin-cam heads. At the same time Chevrolet began to favour the use of the bulky new six-speed ZF gearbox, which meant that the existing engine also began to look too long.

In August 1985, therefore, Lotus were asked to design a completely new lightweight engine – aluminium heads with Nikasil cylinder wall coatings, aluminium block and every other detail – yet they were given no more time to complete the job! New design work on what became known as the LT5 unit concentrated on making this a very compact engine – as low as possible to reduce the width across the vee at the top of the V8 unit, and as short as possible to ease the installation of the ZF gearbox.

Because of this the team chose its own bore and stroke dimensions. By comparison with the standard Chevrolet

Look carefully at the layout of the Lotus Type 909 V8 engine, and this engine – the LT5 V8 produced by Lotus for use in the Chevrolet Corvette ZR-1 – and there are obvious similarities. Although the two engines shared no common components, they were drafted by the same experienced design team.

unit, these were as follows:

Lotus LT5 32-valve : 99mm bore x 93mm stroke
Chevrolet L98 ohv : 101.6mm bore x 88.4mm stroke

The space restrictions were serious, especially as Chevrolet insisted on the existing engine's cylinder bore spacings being retained. One day, they said, the cylinder heads might be used on the cast iron block – of which no fewer than 60 million had already been made! – though in the event that option had not been taken up four years after the design had been unveiled. In fact the new LT5 engine weighed 39lb more than the cast-iron L98; clearly the weight would have been much higher if cast iron blocks had been used.

Lotus therefore had to use every trick in the engine designers' book to keep down the bulk. A narrow cylinder head layout with short valve/spring assemblies was an obvious choice – the angle between the lines of inlet and exhaust valves was only 22deg – along with an 11:1 compression ratio – though it could still run on unleaded 87 octane fuel – but Rudd's engineers never even considered reducing the familiar

Another fine contract job carried out by Lotus in the mid-Eighties was to produce a high-performance 16-valve twin-cam version of the Chrysler Corporation's existing 2.2-litre four-cylinder unit, which was used in many of that Corporation's front-wheel-drive cars.

90deg vee-angle of the cylinder block, which would have reduced the engine's width but increased its height. Pistons which barely cleared crankshaft webs were adopted, and chains rather than cogged belts were used to drive the four overhead camshafts.

Among the many thoughtful details were the use of GM Delco fuel injection in which there was a two-stage, three-valve, throttle body and separate tracts to each of the engine's 16 inlet valves. There was actually a removable 'power key' on the engine. When set to 'Full' fuel was supplied to every inlet track, but at 'Normal' only the primary port valves and injectors operated, which made the engine more docile; the key could be removed to lock the settings in the 'Normal' condition.

Lotus designed the whole engine in a matter of months – a massive job when the detail involved is considered – and the first prototype ran in May 1986. It was, of course, one thing to develop an engine for which the target horsepower was a peak of 400bhp, but it was quite another to make sure it also passed every known – and projected – exhaust emission and noise standard.

As GM was determined that the new engine should also power a Corvette which beat the 'gas-guzzler' tax – this required cars to be more economical than 22.5 US miles-per-gallon in standard Environmental Protection Agency tests – Lotus had a real challenge on their hands. In the end every single requirement was met: the idling speed of what was a complex engine was set at no more than 450rpm, the homologated power output was 375bhp at 5,800rpm and the peak torque 370lb.ft at 4,800rpm.

The new unit, which was only available in the Corvette ZR-1, was unveiled in 1989, with assembly being carried out for GM by Mercury Marine, the world's largest manufacturer of boat engines, at Stillwater, Oklahoma. Once they had completed the job, Lotus were no longer involved in manufacture, which left GM to bask in the use of North America's most powerful production car engine.

By 1992 there was some talk at Hethel of developing a 'super-Esprit' for the mid-Nineties, and we know that after some modifications the LT5 engine could be squeezed into the Esprit's engine bay.

By 1988 the shape of this Dodge Daytona was familiar, but under the skin of this version, the Shelby Z, was a front-wheel-drive package which included the Lotus-designed twin-cam 16-valve unit.

Lotus-Carlton and Lotus-Omega

The 3.6-litre 377bhp-engined GM supersaloon twins, unveiled in 1989 and going on sale in 1990, were among the most successful projects completed by Lotus in the Eighties.

As these were front-engined cars, they are described more fully in Volume 1 to this book.

Millbrook proving grounds

Although General Motors took control of Vauxhall in the mid-Twenties and Opel in the Thirties, they did little to merge the interests of these companies until the Sixties. It seems crazy today, but GM was confident enough to let *both* companies develop their own proving grounds.

At the end of the Sixties, when Vauxhall-Bedford was still operationally separate from Opel, the British company built a vast new proving ground at Millbrook, immediately to the east of the M1 motorway. This enormous 700-acre site was completed by 1970, and not only included 30 miles of test tracks and an impressive number of indoor test facilities, but also a circular banked two-mile high-speed bowl which allowed cars to reach their hands-off top speeds.

After GM merged Vauxhall-Bedford with Opel, giving the German concern European leadership of all passenger car design, Millbrook was very active, but by the Eighties it was only being used by Bedford and several contract customers.

After Lotus had been taken over by General Motors they began using Millbrook as a matter of course. Then, in the expansionary days of the late Eighties, it was decided to buy the facility. In April 1987 it was announced that Group Lotus had paid £7.5 million for the Millbrook site.

Even though Lotus' own activities were drastically reduced from mid-1992, this was an acquisition which looked as if it might pay off in the long run as ever more private clients began to use its facilities. In particular the twin-straight-mile, and the two-mile banked circuit tracks were extremely useful to anyone wishing to establish the flat-out performance of their cars, magazine testers and record breakers all finding these tracks invaluable.

Following the takeover of Lotus by General Motors, Lotus then absorbed the large proving ground at Millbrook, in Bedfordshire, which had originally been built by Vauxhall for their own use. This X180-style Esprit Turbo is pictured on one of the many sinuous curves which make up the hill route.

One of the most astonishing features of the Lotus-owned Millbrook proving ground was the totally circular high-speed circuit, which was wide and smooth. Except for a few miles-per-hour which were inevitably lost due to tyre scrub, any of the world's production cars could reach their top speed there.

Other research and contract work

In the Eighties Lotus carried out a great deal of contract and consultancy work for almost all of the world's car makers, much of which will never be made known to the public. Apart from designing the turbocharged engine installation for Austin-Rover's MG Metro Turbo and building the final twin-plenum engines for the last run of Rover Vitesses, the company prepared an 'Evolution by Lotus' version of the Opel Kadett/Vauxhall Astra GTE 16-valve – which was cancelled – tweaked the suspension of several Japanese cars – including the Toyota Supra, and the Isuzu Piazza – and of

several American cars (models unstated) and improved the ride and handling of cars for many other manufacturers.

Two major projects revealed in the same period were the installation of a pair of semi-anechoic chambers to measure noise and vibration, and at the same time to develop anti-noise technology in that chamber.

What that really meant was to 'fight fire with fire'. An electronic noise-generating system was evolved which could measure noises being produced in a car's cabin, and emit similar noises which would effectively damp out the originals. In effect this meant destroying sound waves of one type by bombarding them with similar waves in a different phase.

Like Active Ride, the problems were severe, but the rewards of a successful programme would be enormous. However, in the straitened financial climate of the early Nineties, was there enough financial muscle to carry it through?

Until the money ran out, the projects ranged from the simple and classic, to the forward-looking. On the one hand there was the SID project, originally based on something looking like an ordinary Esprit, but on the other there was the decision to remanufacture obsolete cylinder heads.

Old-type Lotus-Ford twin-cam cylinder heads finally went back into production after several years of lobbying from enthusiasts and Lotus restoration specialists. This contrasted sharply with Lotus' part in developing an ultra-modern racing bicycle with a monocoque Kevlar frame, which was the machine used to win the 4,000m pursuit gold medal at the Barcelona Olympics in 1992. In happier times Lotus would have been asked to put this bike into production.

The SID (Structural Isolation Dynamics) project was an indication of the way Lotus engineers liked to think, if they had the time, the resources and the backing. Not only did this car use a mid-Eighties type of MG Metro 6R4 rally car powertrain – mid-mounted 3-litre V6 engine and four-wheel drive – but it also incorporated the latest version of Lotus' Active Ride suspension, as well as helping to develop a four-wheel-steering installation and damping out noise and harshness from the suspension.

Perhaps it will never be known how much innovation came from Hethel in this period, nor how much *might* have gone into Lotus cars if the company had still been profitable. If only ...

This is the fabulous racing bicycle developed by Lotus, which Chris Boardman used to win the individual pursuit gold medal in the 1992 Olympic Games at Barcelona. The 'frame' is a carbon-fibre monocoque, with titanium inserts.

CHAPTER 8

Jensen and Talbot

The Lotus engine users

This book would be incomplete if the story of the two other cars which used the 16-valve Lotus engine in the Seventies and early Eighties – the Jensen-Healey and the Talbot Sunbeam-Lotus – were not related. Both were built in considerable numbers and therefore contributed materially to Lotus' financial health during an often troubled period. Without the Jensen-Healey business, Lotus would have been financially overstretched for a couple of years when their new engine facility was ready, but the cars for which it was intended were not.

The Jensen-Healey had its origins in the Sixties and in the formation of British Leyland. Donald Healey had enjoyed a successful business relationship with BMC throughout the Fifties and Sixties in links involving the Austin-Healey marque, but once British Leyland, controlled by Sir Donald Stokes, came into being of which the old BMC was just one part – it became clear that Healey was to be frozen out. Then and there, Donald and his son Geoffrey started to look around for a new opportunity.

The Austin-Healey 3000 – the 'Big Healey', as it was affectionately known – had disappeared by 1968, so the Healeys began to talk to the California-based businessman, Kjell Qvale – who had garage businesses which had sold many Austin-Healeys – about a successor. In 1968, with little idea of where they would build a new car, they began to evolve a new two-seater sportscar, styled originally by Hugo Poole, then modified by Bill Towns, which was to use Vauxhall Viva GT suspension and running-gear.

In 1970, the opportunity came for Qvale – who was a rich

man – to buy Jensen Motors, of West Bromwich, from the merchant bankers William Brandt. This was achieved in April 1970, whereupon Qvale installed Donald Healey as the company's new Chairman and instructed Jensen's Chief Engineer, Kevin Beattie, to take up the new project and productionize it as speedily as possible.

At this point the new car ran into trouble on two counts: styling and choice of engine. The style had to be modified to meet certain legislative requirements and was later altered again to take account of Qvale's taste, the result being a rather anodyne open two-seater shape.

The engine problem was simple – at first. While the Vauxhall Viva GT unit was freely available to Jensen and Healey, it rapidly became clear that the ever-tightening US exhaust emission regulations would strike hard at its power capabilities. For 1972 it seemed certain that Jensen would end up with a gutless wonder.

The search was suddenly on at West Bromwich for alternative power and, with it, an alternative gearbox. Amazingly, nobody seems to have approached Lotus at first, even though it was known that their engine had originally been built up on the same cylinder block as that used in the Vauxhall Viva GT. Colin Chapman, however, made the first approaches, but could only offer up to 60 engines per week; Jensen were planning to build up to 200 cars per week, 60 per cent of which would be destined for sale in North America.

In a great rush, therefore, Jensen investigated the German Ford V6 engine of the Capri RS26000, the 2-litre BMW four-cylinder unit, the Volvo four-cylinder and even the new

Jensen were the first major users of the Type 907 engine, in the two-seater Jensen-Healey sportscar. The Lotus connection was never advertised on the car's exterior.

Wankel rotary engine from Mazda. It was not long, however, before Chapman reopened negotiations on behalf of Lotus when it became clear that the Elite would not be ready for at least two years, which meant that he could supply much higher quantities of the Type 907 engine to West Bromwich.

After Chapman and Fred Bushell had flown to California to finalize the deal with Kjell Qvale, the deed was done in the spring of 1971, with the forecast that Jensen would be taking up to 15,000 engines per year, or 300 per week – the quantities quoted were getting quite out of hand! – and that deliveries would begin early in 1972. The engine, it was stated, would be rated at 140bhp for Europe and 115bhp for the United States. As in the forthcoming Lotus installation, it would be installed at an angle of 45deg in the engine bay, leaning over towards the left side of the car.

The Jensen-Healey – for such was its name, there being no

An early Jensen-Healey on test at Lotus, with the car stationary, but 'running' on the roller-rig chassis dynamometer. This particular car is in left-hand-drive 'Federal' guise – and the Lotus engine had twin Zenith-Stromberg carburettors.

mention of Lotus – was launched in March 1972 in time for showing at the Geneva motor show, and the first deliveries began soon afterwards. It was a neat but rather plain-looking car, for which the drag coefficient was claimed to be 0.42. The Vauxhall suspension and back axle had been retained, though the Lotus engine was backed by a non-overdrive Sunbeam Rapier four-speed all-synchromesh gearbox.

Deliveries began at a total British price of £1,811, which compared with £2,116 for the current Lotus Elan Sprint, and original road tests showed that the car had a top speed of nearly 120mph, which was achieved at 6,650rpm – just over the peak of the power curve – but that it was outpaced and outhandled by the less powerful Elan. It was not known that

Lotus would rate their own engine at 160bhp for the new-generation Lotus cars, and at the time we thought the 140bhp rating of the engine as supplied to Jensen, and the performance it produced, was quite satisfactory.

The story of the Jensen-Healey is easily told, for the car finally went out of production in the spring of 1976 when Kjell Qvale called in the Receiver to close down the unprofitable Jensen business. This was not really the fault of the Jensen-Healey – or the estate car-style Jensen GT, a sort of Elite copy, which arrived in 1975 – though sales never approached the 15,000 per year once mentioned; it was that in the aftermath of the Suez war of 1973, the market for the big Chrysler-engined Jensen Interceptors faded away rapidly

and dragged the company down into bankruptcy.

No major changes were made to the Lotus-supplied engine of the Jensen-Healey in the four years in which it was in production, though the Chrysler gearbox behind it was changed for a German Getrag box from November 1974. A total of 10,912 of these cars were built and a breakdown of statistics follows:

Production by model:	Jensen-Healey	10,453
	Jensen GT	459

Annual production: (Jensen-Healey)		
	1972	705
	1973	3,846
	1974	4,550
	1975	1,301
	1976	51

Deliveries by market: (Jensen-Healey)	Domestic	1,914
	USA/Federal	7,709
	Other export	830

The 'Federal' Jensen-Healey, like later Lotus-used engines, had twin horizontal Zenith-Stromberg carburettors. In this illustration the cam drive belt cover had not yet been fitted, though the unique-to-Jensen tubular exhaust manifold had.

Assembly of engines for the Jensen-Healey under way at Hethel. The lightweight, but very strong cylinder block is much in evidence.

Demand for the Jensen-Healey was already contracting before the company went into liquidation. There is no doubt, however, that the Jensen business was very valuable to Lotus, even though they were left with a large excess engine-building capacity in 1976 when the Jensen-Healey GT model died.

The imbalance between Lotus engine production at its peak and at its lowest point is remarkable. In 1974, 4,550 Jensen-Healeys and 687 Lotus Elites were built, making a total of 5,237 cars. In 1977 total production was 1,070 of all types, and the Jensen business had been lost.

Lotus, no doubt, were delighted to be approached by Chrysler UK for engine supplies in 1978, especially when the company started talking about substantial numbers of engines. Amazingly, however, the Chrysler Sunbeam-Lotus – which was renamed Talbot Sunbeam-Lotus almost as soon as it had been announced – was not a carefully researched and product-planned performance car, but the result of a brainwave by Des O'Dell, Chrysler's dynamic Director of Motorsport. It was only after O'Dell decided on a particular car layout which he considered ideal for rallying that he approached his management for backing.

Jensen-Healey installation of the 2-litre Type 907 engine, complete with Dellorto carburettors and distinctively badged camshaft covers and featuring a large separate AC air cleaner in the engine bay, connected to the engine by flexible trunking.

The original Chrysler Sunbeam-Lotus of spring 1979, painted black and silver, with Lotus roundels on the car's flanks. By the time the car went on sale in the summer it had been rebadged as a Talbot.

Lotus supplied a 2.2-litre version of the 16-valve engine for the Chrysler/Talbot Sunbeam-Lotus, rated at 150bhp. Apart from the 'Chrysler' camshaft covers on this prototype – which reverted to standard 'Lotus' for production cars – and the special carburettor intake trunking, the engine looked very familiar to all Lotus enthusiasts.

There was nothing spartan about the facia and instrumentation of the Talbot Sunbeam-Lotus. There was no 'Lotus' badging in this style – but a driver surely didn't have to be told!

At the end of 1977, Chrysler's 16-valve Avenger-BRM was outlawed from motorsport by a change in regulations, and O'Dell was looking around rather desperately for a new opportunity. For the basic 'chassis', he decided almost at once to use the Chrysler Sunbeam hatchback, which used a shortened version of the Avenger's floorpan and had a smart but simple three-door superstructure.

The problem was that a new and very powerful engine was needed to make the car competitive. At the time, the Ford Escort RS was completely dominant in international rallying, setting the standard with a 240–250bhp 2-litre engine, while the Vauxhall Chevette HS – which used some Lotus components, described below – was almost as fast.

There was no way that Chrysler themselves could produce a competitive engine – time, money and design expertise were all lacking – so O'Dell looked around for a proprietary unit. It needed little investigation to show that the 16-valve Lotus engine was the only suitable one. Early in 1978, therefore, O'Dell acquired a rally-tuned Type 907 from Lotus, shoehorned it into a Sunbeam hatchback, matched it to a

The Sunbeam-Lotus in its element – with a works car rallying on a variety of surfaces. This was Henri Toivonen and Paul White on their way to winning the 1980 Lombard-RAC Rally outright. Other works-supported cars took third and fourth places in the same event. The power output of a works car, with their own big-valve engine, was around 250bhp.

Works competition Sunbeam-Lotus cars under preparation at Coventry in 1981. With a total of only six cars, Talbot won the World Rally Championship for Makes in that season.

This is a Lotus publicity shot, of course, for no ordinary Sunbeam-Lotus was normally badged with 'Essex' on the doors. But was the hatchback *really* so much taller than the Esprit Turbo?

five-speed German ZF gearbox and a more robust back axle, and went testing. The 'motor industry grapevine' system worked well, for O'Dell's Development Engineer, Wynn Mitchell, had known Lotus' Mike Kimberley for years.

On the strength of one car, some test results and an outstanding – unhomologated – test debut on the Mille Pistes rally, when Tony Pond took second place, O'Dell won approval for a run of cars to allow him to homologate the Chrysler Sunbeam-Lotus. To get the car approved into FISA Group 4 – in which the Ford, Vauxhall, Fiat 131 Abarth and other supercars were all placed – 400 would have to be built in less than a year. Chrysler, however, made hasty inquiries around their dealers and overseas concessionaires and decided they could do much better than this. By the time the car was announced they had laid plans to build no fewer than 4,500 cars!

Between the decision to build the cars, which was taken in mid-1978, and the public launch of the machine in March 1979, there had also been a corporate upheaval, though this never affected the Sunbeam-Lotus' career. Chrysler of Detroit tired of sustaining losses in their European subsidiaries and sold out to the Peugeot-Citroen Group in September 1978. As part of the rationalization and realignment programme which followed, the British Chryslers of 1978 became Talbots from August 1979.

This car, in fact, was the first to use the enlarged 2.2-litre Type 911 Lotus engine which, compared with the Type 912 fitted to S2.2 Lotuses, had different carburettor settings, ignition, lubrication system and main bearing castings, and a different oil sump to clear the front-suspension crossmember.

For the Talbot, the engine was rated at 150bhp (DIN) at 5,750rpm, while peak torque was 150lb.ft at 4,500rpm,

Apart from the rollcage inside the passenger compartment and the tailgate spoiler, this Talbot Horizon looks relatively normal ...

... but under the skin there was a mid-mounted Esprit Turbo engine and Citroen-type transmission. It was Coventry's first thoughts on the theme of a 200-off Group B rally car for the mid-Eighties, later overtaken by the Peugeot-designed Type 205 rally car. Only one example was built.

Among their many engineering contract activities, Lotus also found time to build this mid-engined rally prototype for Citroen. The basic car was Visa, but behind the seats there was a 2.5-litre Citroen Reflex engine of about 220bhp.

which effectively put it halfway between the engines once supplied to Jensen and those used by Lotus themselves. In its class, of course, this made the Talbot Sunbeam-Lotus very competitive in road-car form, for the limited-production Vauxhall Chevette HS had 135bhp, while the Ford Escort RS1800 (with 120bhp) had recently been dropped.

The five-speed ZF gearbox fitted to the Sunbeam-Lotus was certainly not as smooth nor as refined as the Getrag unit shortly to be standardized on S2.2 Lotus Elites and Eclats, but it was very strong indeed, and could be modified in many ways to make it ideal for rallying in all conditions. Fifth gear in the road car application was a geared-up overdrive, and as a result the car was almost exactly as fast in fourth as in fifth gear. *Autocar*'s testers recorded 121mph, with 0–60mph in 7.4sec, and a standing quarter-mile time of 15.6sec.

Incidentally, although there was no external evidence of a Lotus involvement in the Jensen-Healey, on the Talbot Sunbeam-Lotus not only was the name 'Lotus' in the car's title, but the famous Lotus 'ACBC' roundels were applied as transfers to the side of the bodyshells, ahead of the doors.

Lotus, in fact, were much more closely involved in the Sunbeam-Lotus project than they had ever been with the Jensen-Healey. Not only did they carry out a whirlwind development and proving exercise on behalf of the Coventry-based Engineering Division of Talbot – *nee* Chrysler – but they were also responsible for part-assembly of all the production cars.

Because of the proposed number of Sunbeam-Lotus cars to built, there was insufficient space for all the work to be done at Hethel. Accordingly, although the engines were machined, assembled and checked out there, car assembly was completed at another building – already owned by Lotus – on Ludham airfield, about 10 miles north east of Norwich and nearly 20 miles from Hethel, in the heart of the Norfolk Broads area.

About 16 Lotus staff were drafted to Ludham, where pre-production began in March 1979 and continued to May 1981. Cars were built up to what I call the 'rolling chassis' stage at Linwood factory a few miles west of Glasgow, where all other Sunbeam hatchbacks were being built – and then transported to Ludham, where the Lotus staff not only mated Type 911 engines to German ZF gearboxes, but fitted these to the cars, along with larger-capacity water radiators and modifications to the transmission tunnel. The completed cars were then sent to the Talbot factory in Coventry for final checking and inspection before being delivered to the dealers.

Sales of the Sunbeam-Lotus began in the summer of 1979, and there were stocks in hand when the last cars were assembled in 1981. Talbot, like many other car makers before them, had overestimated the demand for what was, after all, a homologation special, and total production in fact was only 1,150 right-hand-drive and 1,148 left-hand-drive cars, which made up a total of 2,298 in all – half of the intended sanction. Most were black, though the last few hundred were an attractive shade of blue, and it is worth recalling that no fewer than 30 examples were supplied to British police forces as well!

After Talbot withdrew from works rallying with these cars, sales virtually stopped, such that the last 150 examples were converted in 1983 to a super-luxury interior specification by Ladbroke Avon Coachworks, in Warwickshire. Assembly had stopped when the entire Talbot Sunbeam range was dropped and the Linwood factory was closed down.

The Sunbeam-Lotus was a very successful works rally car, even though it had a tentative and unlucky start. In 1979, Tony Pond was hired to drive the factory-built machines, but

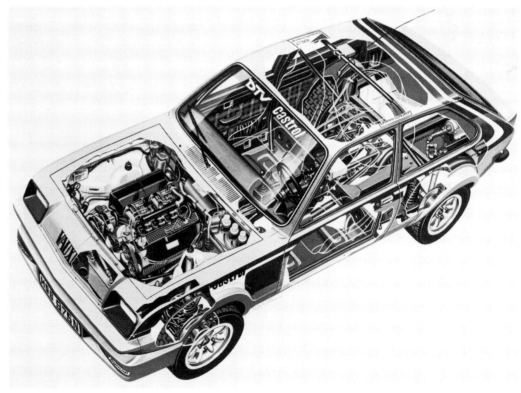

In 1976, Vauxhall produced and homologated the Chevette HS, equipped with their own 2.3-litre engine bottom end, to which a complete Lotus 16-valve Type 907 head, valve gear and carburation had been grafted. It won its first international rally in mid-1977.

achieved very little in spite of leading the Scottish and Manx rallies and holding third place on the Lombard-RAC until the last night of the event, when he crashed and had to retire.

Pond was no longer a team member in 1980, but the late Henri Toivonen and Guy Frequelin (Finnish and French, respectively) made up a two-car driving team. There was a gradual build-up of success which culminated in Toivonen winning the prestigious Lombard-RAC rally outright, with two other Sunbeam-Lotuses (Frequelin and Russell Brookes) in third and fourth places.

The truly memorable season, however, unfolded in 1981 when the Talbot factory team, using a mere six cars, tackled the entire World Rally Championship series. Although they won just one event — Guy Frequelin finished first in the Argentinian Codasur event — they also took second place on five more occasions — Monte Carlo, Portugal, Corsica, Brazil and San Remo — which was sufficient for them to win the Championship for Makes. Guy Frequelin himself was second in the Driver's series behind Ari Vatanen, who used Ford Escorts throughout the year.

At the end of that season, Talbot withdrew from World Championship events, while the parent company, Peugeot, began the design of a new four-wheel-drive mid-engined car for 1983 and beyond. In the meantime, however, Des O'Dell and Lotus had co-operated in the building of a mid-engined Talbot Horizon prototype, where the engine was placed behind the front seats, driving the rear wheels only. This might, indeed, have become Talbot's new Group B rally car if the arrival of the Audi Quattro had not convinced most rival manufacturers that they, too, would need four-wheel drive in the future. But wouldn't a turbocharged, mid-engined, Horizon lookalike have been exciting?

Finally, mention must also be made of Vauxhall's attempts to get on terms with Ford in the world of international rallying, and the way in which Lotus were indirectly involved. Vauxhall first used the Magnum Coupes and then progressed to Chevette HS and HSR models.

Dealer Team Vauxhall began by using Magnum Coupes fitted with the conventional single-cam iron-block 2,279cc Vauxhall engine. Next, however, they spotted a new regulation which allowed the use of an alternative cylinder head in

competitions if they could guarantee that 100 kits had been sold for conversion purposes.

No-one now believes that anything like 100 such heads were ever supplied, but somehow Vauxhall achieved recognition and were able to go rallying with Magnums fitted with Lotus Type 907 cylinder heads, camshafts and carburettors mated to the original Vauxhall cast-iron cylinder blocks. Suddenly, it was 1968 all over again.

The Magnums were never very successful because they

The Chevette HS rally cars used Lotus 16-valve heads and carburation until the spring of 1978, when homologation was rescinded. To give clearance for the left-hand-drive steering column (for Pentii Airikkala's cars) the engine was installed at 35deg, rather than 45deg to the vertical. To disguise the head's parentage, the cars had 'Blydenstein' – after the company who prepared them – on the camshaft covers.

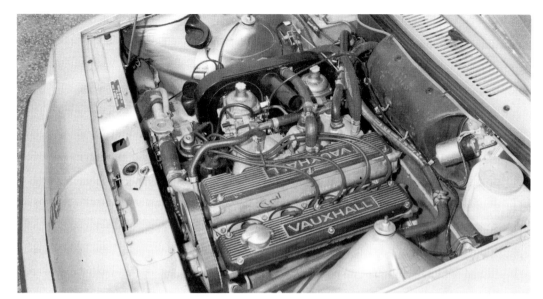

Chevette HS production cars which went on sale in 1978 had Vauxhall's own 16-valve cylinder head, an entirely different casting from the Lotus unit, for the included angle between valves was altered and the camshaft covers were also at a different angle.

This is what the Chevette HS road car looked like – some of these cars possibly being fitted with Lotus engines later, as Vauxhall parts supplies had long since run out.

were too bulky and had unsatisfactory traction and handling, but the Chevette HS which followed was a much more competitive and serious project.

DTV persuaded Vauxhall that the only way to match and eventually beat the all-conquering Escorts was to produce the Chevette HS, for which 400 examples had to be built to secure Group 4 homologation. The HS models, effectively, were mainstream Chevette hatchbacks with the 2.3-litre slant-four Vauxhall engine and a *Vauxhall*-designed 16-valve cylinder head allied to a stronger gearbox, and it was this specification which was planned to go into limited production.

Somehow, though, the Chevette HS came to be homologated in November 1976 on the strength of a single car – *the* prototype. This was well before production cars could possibly be built, and it was fitted with the Lotus Type 907 cylinder head! In this form it was easily capable of developing 240bhp, and within six months the Chevette HS had started to win British international events.

Soon there was something of a rumpus in the rallying world, not only because the Chevette had been homologated on very flimsy evidence indeed, but because it was so clearly meant to drive a great wedge into the spirit of rallying regulations.

The Chevette HS production car was not actually put on sale until April 1978 – 16 months *after* the car had originally been homologated – when it was seen to have the Vauxhall, rather than the Lotus cylinder head, the two light-alloy castings being entirely different. Predictably, therefore, there was a further sporting storm, the result being that the authorities temporarily rescinded the car's homologation, only being persuaded to restore it when the Vauxhall cylinder head and other proper production details had been standardized in the paperwork.

Chevettes with the Lotus-Vauxhall engine had always been fast, reliable and competitive, and certainly this was one of the factors which inspired Chrysler's Des O'Dell to produce a similar car for his own use. Incidentally, all credit to Vauxhall's engineers for producing a cylinder head which was eventually persuaded to liberate as much power as the race-proved Lotus head had always done, and to DTV's Bill Blydenstein for persisting against all the odds. The Vauxhall head, incidentally, had its valves opposed at an included angle of 31deg, compared with 38deg in the Lotus head.

CHAPTER 9

The DeLorean connection

Involvement in a lost cause

Mention of the ill-fated DeLorean project must be made in this book for two important reasons: one is that Lotus engineers were responsible for most of the design and development work on the car which went into production in Belfast's Dunmurry factory; and the other is that there were many close technical similarities in the chassis layout of the rear-engined DeLorean and the mid-engined Lotus Esprit. This is certainly not the place, however, to go into all the sordid commercial details surrounding the DeLorean company – which have already been published in several quarters – but merely to outline how Lotus came to be involved in the project and the part they played.

John Zachary DeLorean had been one of General Motors' most publicity-conscious 'young lions' in the Sixties and early Seventies. However, he abruptly resigned from a very senior corporate position in 1973, in circumstances which have never properly been explained, and within a year he had started to think about a new car design of his own. The DeLorean Motor Company was founded in 1974, by which time the concept of a rear-engined – *not* mid-engined, please note – two-seater coupe had already been dreamed up. It was to have stainless-steel skin panels, but a structure which mainly consisted of composite materials some way removed from simple glassfibre; right from the start, too, it was always scheduled to have lift-up gull-wing doors. Only two production sportscars in the world – the Mercedes-Benz 300SL of the Fifties and the Bricklin SV-1 of the Seventies – had ever had such doors and neither car had sold in large numbers. At the end of 1974, DeLorean had picked Giorgetto Giugiaro and Ital Design to shape his dream car, and perhaps it was at this point that his first links, however tenuous and unconscious, were made with Lotus, for the talented Italian had just finalized the styling of the Esprit.

The first prototype, complete with a transversely-mounted Citroen CX four-cylinder engine and transmission, was completed in 1976, but it was found to be seriously underpowered. The second prototype of 1977 retained the same basic Ital Design body style and structure, but was equipped with the single-overhead-camshaft 90deg V6 2.66-litre Renault 30 engine – the PRV unit also used by Peugeot and Renault – and the transmission as also fitted to the Renault 30. In the Renault installation, incidentally, the engine was ahead of the final drive and the transmission behind it, driving the front wheels, whereas in the DeLorean the engine was behind the final drive and the transmission in front of it; to ensure correct operation in the DeLorean it was essential for Renault to flip over the crownwheel-and-pinion set, like the Alpine-Renault A310 V6 installation.

The project had already been dragging on for more than four years before DeLorean found anywhere to build production cars, and the state of the Renault-engined prototype could best be described as 'elementary'. The original scheme had been to use a multi-section Elastic Reservoir Moulding (ERM) plastics monocoque, in which 0.5in thick sheets of open-cell urethane foam were sandwiched between sheets of glassfibre – and then moulded into shape under pressure in simple press tools. There were front and rear steel subframes to support suspension, engine

The ill-fated DeLorean coupe on which Lotus development engineers did such a magnificent job between 1978 and 1980. The company's failure was certainly nothing to do with Lotus.

and transmission components, all bolted up to the plastics shell. However, even before the second prototype had been built, the ERM concept had been abandoned for the time being and pure glassfibre was used in its place. The very first designer to use a glassfibre monocoque was Colin Chapman, for the original Elite of 1957.

In 1978, John DeLorean's three major problems were to raise finance to have the car put into production, to find a production site, and to have the design refined and finalized. It is now a matter of history that he nearly settled on a site in Puerto Rico, that a site near Shannon Airport in the Republic of Ireland was also considered, but that he eventually persuaded the British Government to lend and grant many millions of pounds for the sports coupe to be built in a brand-new factory at Dunmurry, just south of Belfast, in Northern Ireland.

DeLorean first invited Lotus to have a look at the car during the summer of 1978, the result being that Colin Chapman and

Mike Kimberley flew out to see the second prototype in Phoenix, Arizona. They have since been quoted as thinking that the car was 'abominable' at that time, and that DeLorean's development chief, Bill Collins, agreed with them.

Even so, although British Government finance for the project was agreed and made public in August 1978, Lotus were not contracted to develop the car until November of that year. For three months John DeLorean had been shopping around in Europe, particularly at Porsche, who had wanted a great deal of money and up to four years to do the job. Consequently, when Lotus agreed to do the same project for less, and in a mere 18 months, they were signed up.

It was a very demanding contract, as Mike Kimberley recalls: 'We started with only a package – a style – and we had to provide engineering facilities – design, engineering development and prototype build. We did it under the guidance of Chuck Bennington, their Managing Director, who spent four days out of every seven at Hethel. We were

Dramatis personae – John DeLorean (centre), Mike Kimberley (right) and Lotus engineering chief Colin Spooner in discussion over the new DeLorean project.

given 18 months to do the job, a pace which no other company in Europe could have achieved.'

The fact that the finance for this business came to Lotus via a Panama-registered company called General Products Development Services Inc (GPD), and not direct from DeLorean, was a puzzling financial wrinkle which has no part of this technical story. There was no doubt, however, that the money was for real, and available. DeLorean established an engineering and purchasing facility in Coventry, and also drafted a small team into Hethel and Ketteringham Hall. Lotus allocated the largest building on the site – a vast hangar which had to be refurbished – for this, their largest-yet consulting job. Lotus demanded, and achieved, virtually a free hand to redesign the car as their engineering judgment dictated. It was clear, right at the start, that the prototype's

unique construction was going to be controversial – as Jeffrey Daniels of *Autocar* pointed out in his technical analysis of October 1977: '... instead of a plastic body on a steel frame, we see in effect a steel body on a plastic frame ...'

Lotus were most unhappy about it all, not merely because the prototype's roadholding and general rigidity were not helped by the big cut-outs in the shell dictated by the gull-wing doors, but also because of the safety limitations. Their own vast experience of both glassfibre-based bodies and backbone chassis-frames led them to surmise that in its original guise the DeLorean would fare very badly indeed in a barrier crash test. Very shortly Grumman – the US aerospace concern – were invited in by DeLorean to analyze the structure on their most sophisticated computers; they did so and predicted that the rear-mounted engine and gearbox

The bare bones of an early DeLorean development chassis-frame, showing clearly how its design philosophy was very much that of the modern-generation Lotus cars.

A display example of the DeLorean DMC12's production chassis, showing how the fuel tank sat between the front forks of the backbone, and the engine was hung out at the back. In many ways, including the wishbone front suspension, independent rear suspension layout and the use of Goodyear NCT tyres, the DeLorean was close to the Esprit – but Colin Chapman would never have considered using a rear-engined layout.

The suspension geometry looks rather like that of the Esprit S3/Turbo cars, but the car is the DeLorean, with the rear-end dominated by the big 90deg 2.85-litre PRV V6 engine.

would catch up with the occupants in the front footwells in a 26mph crash test!

Before the DeLorean was ready to go into production it had virtually become a rear-engined (Renault-engined) Lotus, with a familiar type of backbone frame and independent suspension inside the Ital Design shape. Even the styling was altered, for in 1979, in mid-contract, DeLorean decided that it had been around too long and needed to be freshened up before going on sale, and he asked Ital Design to facelift everything. As Lotus were already on such a tight schedule that they had finalized the skin contours well before the rest of the car was fixed, it meant that new skins, window shapes and cut-outs had to be redrawn, and that everything abutting those profiles – which meant the composite bodyshell, trim panels, glass and many other details – had to be modified. The

change, Mike Kimberley says, came 10 months into the 18-month contract and caused a considerable delay, which stretched the job, in the end, to 25 months. The project, which was absorbing well over half of Lotus' entire strength when at its peak, would have been finished in May 1980, but was eventually wound up at the end of the year. As has already been pointed out, the sheer size of this project inevitably meant that there was delay to the development and production of Lotus' new models for the Eighties, although much of this leeway was subsequently made up.

The DeLorean sports coupe went on sale during 1981, but the company collapsed into bankruptcy in 1982, and this meant that the car was only ever marketed in the United States. In Federal form its PRV V6 engine had been enlarged to 2,849cc – as in the latest Volvo 760GLE saloon, for

The front suspension of the DeLorean sportscar had clear historical links with that of the Esprit, and it had the same sort of job to do. The fuel tank is low down between the front arms of the backbone frame.

instance – and it produced up to 130bhp. The unladen weight was only 2,840lb, which reflected great credit on the Lotus design team and their experience with such structures. It was not, however, anything like as quick as its looks suggested. *Road & Track*'s authoritative road test showed that the top speed was a mere 109mph – in fifth or fourth gear – that 0–60mph acceleration took a rather pedestrian 10.5sec, and that 0–100mph needed no less than 40sec and a lot of straight highway. All this helped to give the car its 'gutless' reputation, right from the start. Day-to-day fuel consumption was quoted as 19mpg (US gallons)/22.8mpg (Imperial gallons).

Apart from the DeLorean's rather limited performance,

there was also the question of its handling. John DeLorean was once quoted as saying that the car handled better than most other sportscars and that the rear-engined layout had to be right, for that was how Porsche built the 911. Since no British journalist was allowed to test the car while it was still in production, it was not easy to find out the truth of the matter.

However, my colleagues at *Autocar* got their hands on a 'clearance sale' example at the end of 1982, and Michael Scarlett had some interesting observations to make, the most penetrating of which was: 'Later, doubts returned with the uneasy, unfamiliarly spooky feeling of a well-overhung engine behind. No matter how good the suspension or the Lotus

DeLoreans in series production at the Dunmurry plant, south of Belfast, early in 1981, when hopes were high and up to 40 cars per day were being built.

development engineer, you can't disguise the effects of a high rear-based polar moment, whose beginnings lead inexorably to earthier moments.'

Everyone agreed that Lotus engineers had done a fine job, starting from a very difficult imposed design layout. The backbone chassis-frame, so like that of the Esprit in general layout, was built for DeLorean by GKN in the West Midlands, and was coated with fusion-bonded epoxy resin for anti-corrosion protection. All the running-gear was neatly packaged around it, not only with the engine and transmission in the extreme tail, but with the fuel tank in the nose, between the front forks of the backbone. Front suspension – by unequal-length wishbones, coil springs and an anti-rollbar, linked to rack-and-pinion steering without power-assistance – and rear suspension – by upper and lower links, a long fabricated semi-trailing arm and coil spring/damper units – were both like that latest Esprit installation.

The rear engine ensured that a majority of weight would be over the back wheels – the actual bias was 35/65 per cent front-to-rear – and to tame this Lotus had provided 195/60HR-14in tyres on 6in rims at the front, but 235/60HR-15in tyres on 8in rims at the rear. The philosophy which went so far to making the Esprit one of the best handling cars in the world had been thoroughly applied to the DeLorean project.

Even so, the result was not entirely satisfactory. Michael Scarlett summed it all up very well with this comment: 'If there were European DMC 12s on sale, I can't see why it should have succeeded one whit more easily here than it did in the easier, sometimes more gullible, American sportscar market. Regardless of John DeLorean's good and bad points, I think now, as I have always thought, simply that it was silly to build and try to sell a rear-engined sports car.'

In many respects, the DeLorean was even made in the same way as the modern Lotus models. The bodyshell was built from two major mouldings by the patented Lotus VARI resin-injection method. John DeLorean had been thinking big, for there were half-a-dozen sets of mould tools – at £25,000–£30,000 each – whose capacity was way above any volume that even the super-optimistic DeLorean had ever forecast.

The failure, and most of the facts of the DeLorean business, are now well known. The car, indeed, had a short and very turbulent life. Pilot production began in December 1980 – as soon as Lotus' job was finished – and the first real production car rolled out on January 21, 1981. The original shipment was made to the USA in June 1981, at which point 40 cars per day were built. This rate was doubled to 80 per day in November, but by this time the car's Stateside reputation was already in trouble. Sales slumped, stocks of unsold cars piled up in both Belfast and North America, and the Receivers, Cork Gully, were appointed in February 1982. Production continued slowly and haltingly for some months, but by October it was all over.

The basic reason was not only that the car, as a driving machine, was disappointing, but that it was also far too expensive at $25,000. I simply do not understand how John DeLorean, with a lifetime of experience in that most down-to-earth corporation, General Motors, could ever have expected to sell 25,000 expensive DeLoreans per year – which equated to up to 600 cars per week – as a result of which the financial break-even point was 10,000 car per year. The reason for the company's bankruptcy, therefore, can be spelt out in one sentence: in 23 months a mere 8,000 cars were built and at the end of the day a proportion of these cars were still unsold.

Lotus, fortunately, were paid for most of their work before the various and ever-evolving DeLorean financial scandals broke out. By that time the story of the DeLorean car was at an end. In the decade which followed, investigations revealed that there had been a colossal amount of financial chicanery. Many millions of pounds of British Government aid never reached Lotus, Colin Chapman took the secret of his share to the grave, Fred Bushell was finally tried, convicted and jailed for his part in the affair, and although John DeLorean was also indicted, he chose to stay in the USA, from where he could avoid extradition to Great Britain.

As far as classic car enthusiasts are concerned, the years have not been kind to the DMC-12, for in the early Nineties a 10-year-old example was worth no more than, say, an Esprit Turbo of the same age, or a Europa Special of the Seventies.

CHAPTER 10

Buying and running an Esprit

Maintenance and restoration

Although the only major problems relating to maintenance and restoration of a modern Lotus car in the early Nineties were the costs involved, all the signs suggested that this situation would deteriorate considerably in the years ahead. There was a very real chance that stocks of obsolete parts, when exhausted, would not be reordered.

Following the cancellation of the entire front-wheel-drive M100 Elan project in 1992, Lotus restoration experts took a gloomy view of the company's survival prospects. With Lotus due to build so few cars from this time onwards – none of which would be front-engined – there was the prospect of a rapid contraction of the business. Not even a company as large as General Motors could be expected to support the indefinite use of a factory complex which had become far too large, and whose surviving money was tied up in huge stocks of parts.

However, it could be assumed that the much-changed company which survived into the Nineties would continue to support the cars which had sold in the Seventies and the Eighties, and this chapter is meant to give practical help to keep those cars up and running into the 21st century.

What to buy

Because this book covers every mid-engined Lotus built since 1976, the advice is separated into sections. As everyone knows, there were two very different model families built in that period, along with many derivatives of each type.

Those involved with the classic car hobby for at least 20 years have seen the most amazing changes in notional values;

these swings apply to all types of car, not merely a Lotus. People who bought classic cars in the late Eighties, in some cases hoping they would be investments, saw their values plummet in the next few years, and thus need no reminding that values can change very rapidly.

In a book like this, therefore, it makes no sense to quote absolute price levels – especially as those which exist in the UK may not bear any relation to those in other countries. Even so, there are lessons to be learned from the *relative* prices asked of different Esprit models:

** Cars with 2.2-litre engines seem to be more desirable than those with 2-litre engines. As this book has already made clear, the larger-engined types are more flexible, and easier to drive without frequent gear changing – and this is reflected in the premium prices commanded by the larger-engined types.

** The Esprit became better and more desirable every time the model was changed, and this is reflected in the value of cars today. A Series 2 is worth more than an original type, a Series 3 more valuable than that, and a turbocharged version more valuable than a normally-aspirated type.

The customer has to pay for an improved product – but it is worth it.

** The Esprit Turbo is worth maybe 50 per cent more than an equivalent normally-aspirated model. It is not only the engine performance, but the improved chassis performance, which justifies this.

** X180-type Esprits introduced in the late Eighties were sure to lose much of their worth in their first 10 years, so *no-one* should buy one new or nearly-new in the hope of making a profit, or of it even maintaining its value.

Parts supply

Following the collapse of the Elan project in 1992, the Lotus dealer chain went into shock, and was contracting steadily during the time that this book was being compiled.

In the UK, for instance, by the beginning of 1992 the franchise network had expanded to cover 29 outlets supplying new cars, along with six Service dealers who could work on existing cars, restore old ones, but not supply new examples. By the end of that year there were only three Service dealers left; at least 10 franchise dealers had abandoned the marque and returned to concentrate on the other cars with which they had also kept their premises busy, and more defections were likely.

The situation in the USA and other territories was similar, and the implications were that Lotus enthusiasts were having to search further afield and travel greater distances to get the parts and the expertise which they needed.

Some of these ex-Lotus dealers were very bitter about the collapse of their market, as Lotus had encouraged them all to think that the front-wheel-drive Elan was going to be their passport to eternal prosperity. Even if they were still allowed to buy parts from Lotus to keep faith with existing customers – which, usually, they were not – in many cases their reaction has been to get as far away from Lotus as possible.

Some 'independent Lotus experts' have recently appeared, making sure that their charges for Lotus work are appropriately high to make up for the uncertain nature of this business. In all cases these concerns have to buy their genuine Lotus parts from a dealer in any case, so an extra mark-up is inevitable.

Providing Lotus are still in business as the Nineties progress, the acknowledged specialists feel sure that parts will continue to be available. Lotus' stated policy is to keep parts and facilities ongoing as long as it is feasible, for which read 'profitable'. Certainly, on the evidence of the company's efforts for older models – Europa and Elan/Elan Plus Two types of the Sixties – this is a very genuine commitment. However, certain non-vital parts for the Esprit have already disappeared and there seems to be no chance of reviving them.

The fact is that supplying parts for obsolete models has always been a very profitable business for Lotus, but it has also meant committing a lot of money to stocking those parts. In the early Nineties, as General Motors reeled under the large losses being notched up by Lotus, the accountants wanted to reduce that financial exposure, and the portents for future supplies are not encouraging.

On the other hand, shortly after the closure of the Elan project there was fear that the entire Lotus car business would soon be closed down. One long-serving and realistic Lotus specialist suggested that mid-1995 might be the closure date because that is when the last new-car warranty for the front-drive Elan was due to expire.

There is a further problem which may eventually have to be faced. It is one thing for a Lotus specialist to manufacture parts for older models, which were relatively simple, but it would be quite another for the same specialist to have to remanufacture parts for complex modern Esprits which have ABS brakes, Japanese fuel injection and complex electronics. Not only that but: 'There are too many variations on the modern types, and they are too complicated' – and this certainly applies to the modern Esprit, as is confirmed in the appendices.

A great deal of expertise, and a good stockpile of parts, for the famous 16-valve engine is still available, and for the foreseeable future I anticipate no problems – except those of paying for the work to be done – in restoring such engines to health. The proprietary parts – Dellorto carburettors, for instance, and Garrett AiResearch turbochargers – are off-the-shelf from non-Lotus outlets, and because Lotus always made much of the engines themselves obsolete parts have been remanufactured from time to time.

The problem of rebuilding gearboxes, or of exchanging transmissions, was getting progressively worse during this book's compilation, and may become critical by the late Nineties.

Although Citroen honoured their promise to keep supplies of five-speed transmissions available while the original type of Esprit was in production, they can surely never have expected the model to stay in production for such a long time. By the time the Citroen-gearboxed Esprit was finally replaced by the Renault-gearboxed model in 1988, all other cars which had used that Citroen gearbox had been out of production for years.

Citroen washed their hands of this design in 1991 when they offered all remaining gearboxes, tools and parts to Lotus, but these were refused. Citroen apparently scrapped their own stocks at this point, which means that when Lotus' own stocks are exhausted no more new supplies will ever be available. This is a potential tragedy for Esprit owners, especially as this gearbox is at the limit of its torque capacity when matched to the turbocharged Lotus engine.

As with the Renault-gearboxed Europas of the Sixties, it is now becoming increasingly difficult to restore such a gearbox, which has meant that the value of good secondhand gearboxes has soared, along with the cost of repairs.

As far as the X180 Esprit is concerned, there should be no problem in getting Renault-made gearboxes – which were more robust than the Citroen type – axles or their components for some years, *so long as Renault continues to make them for their own purposes*. Lotus has never, and will never, manufacture parts for such transmissions – they merely stocked parts supplied by the original manufacturer. One can only hope that Renault will continue to service their own cars for which these gearboxes were originally designed, for at least as long as there is a demand from Lotus owners.

Incidentally, even if anyone would consider making an Esprit non-original, it is not practical to try to update a Citroen-boxed Esprit by fitting the Renault GTA gearbox instead. The reason is not that the new gearbox will not fit – it can be made to, if the chassis is changed! – but because the Renault gearbox is incompatible with inboard rear disc brakes. Such a conversion, therefore, would involve a wholesale rebuild. The cost is prohibitively high.

In the early Nineties, however, supply and availability of Lotus-made parts for all modern mid-engined Lotus models was still very good. In particular body sections were freely available, and expert Lotus restorers know that Lotus have historically been very flexible about the size of new mouldings which could be supplied for repairs on a one-off basis. Everyone assumes that moulds would never be destroyed in any case, but would eventually be sold off to a business which

was already an expert in working GRP.

For the restorer, the problem with making individual (replica) body sections is that although they might look right, these cannot be exactly the same as those *originally* moulded at Lotus by the VARI process. On the other hand, it will always be possible to repair battered panels or patch into existing bodyshells.

Parts Lists, Service Manuals, Handbooks and any other literature which was published for these models are theoretically still available, though shortages occur from time to time. Many have been reprinted, at the insistence of the dealers. Lotus owners are advised to inquire of Lotus dealers – especially Service dealers – to find out what was originally published. Even though they are expensive, they repay their investment many times by making clear what is actually involved.

Repairs and rebuilds

While almost all the parts are available, in the case of an unfit chassis I have been advised that it is often preferable to replace, rather than repair that component. Although Esprit and Excel generation chassis were more robust than those designed for Sixties-type Lotuses, there was always a tendency for a bang on one corner to 'ripple' down the backbone to cause distortion elsewhere.

Not only is the chassis-frame of modern Esprits galvanized – and a repair or restoration which involves heating up and reshaping the frame may, in addition to destroying the original galvanic protection, release noxious fumes in the process – but it is a complex item to true up and align to within the precise limits that a Lotus deserves.

Unless the chassis-frame of a modern Lotus is somehow damaged – usually in an accident – it rarely corrodes. Lotus' decision in the Eighties to give an eight-year anti-corrosion warranty on these frames was completely justified by experience, and it now means that a classic Lotus owner should usually be confident of the condition of the frame.

Because deterioration of the bodyshells on all these models is virtually non-existent – gel cracks are virtually a thing of the past, and there is no 'corrosion' as such of a GRP-based body panel – the demand for new bodyshells is virtually nil.

Compared with an earlier type of Sixties Lotus shell, the Eighties shells seem to last much longer, and even when old they need very little structural repair to bring them back to as-new condition.

As a modern Lotus gets older new panels can either be repaired by conventional methods, or by moulding replacement sections, or 'corners', into place.

Clubs

For a time in the late Eighties and early Nineties there were three Lotus one-make clubs which were rivals: one factory-backed, another a sizeable independent operation, and the third a small Midlands-based organization. Now that the official factory-backed Club Team Lotus has been wound up, and the Lotus Drivers' Club has not expanded, the independent club – Club Lotus – is the most significant organization, with a large and growing membership.

Club Lotus, as run by ex-Lotus sales director Graham Arnold, claims 7,000 members worldwide, and has always been completely independent of the Lotus factory. This is essentially a practically-biased, rather than sporting-orientated organization, which exists to help owners of *all* Lotus models, modern and old, to keep their machines on the road, and not only covers parts supply and technical advice,

along with special insurance schemes, but holds repair and restoration seminars, parts fairs and organizes race circuit 'test days' – which means that Lotus owners can have a blast round a circuit under almost controlled conditions without actually being in a motor race.

It has issued a large number of technical bulletins covering different models, restoration topics, advice on laying up for the winter *and* recommissioning the car for a further season.

The club is to be found at:

Club Lotus
PO Box No 8
DEREHAM
Norfolk NR19 1TF
UK

Tel: 0362 694459
Fax: 0362 695522

Now that *Team Lotus World* – as edited by Andrew Ferguson of Club Team Lotus – is no more, the only publication to concentrate solely on the Lotus marque is a British-based magazine, born in 1992, which is called *Lotus Elite*.

APPENDIX A

Technical specifications

Over the years, Lotus has had to produce several different engine specifications to satisfy local regulations for a particular market. To quote them all would be confusing. Accordingly, I have concentrated on the two main markets, Britain and the USA.

Esprit S1 – produced 1976–8
Engine: 4-cyl, twin-overhead-camshaft, 95.28 x 69.24mm, 1,973cc, CR 9.5:1, 2 twin-choke Dellorto DHLA carbs. 160bhp (DIN) at 6,200rpm. Maximum torque 140lb.ft at 4,900rpm.
Transmission: 5-speed all-synchromesh Citroen SM/Maserati Merak manual gearbox, in unit with engine and final drive. Final drive ratio 4.375:1. Overall gear ratios 3.33. 4.24. 5.77, 8.49, 12.77, reverse 15.14:1. 21.85mph/1,000rpm in top gear.
Suspension and brakes: Ifs, coil springs, wishbones, anti-rollbar and telescopic dampers; irs, coil springs, lower wishbones/semi-trailing arms, fixed-length driveshafts and telescopic dampers. Rack-and-pinion steering, no power-assistance. 9.7in front disc brakes, 10.6in rear disc brakes, with vacuum servo assistance. 205/70HR-14in front tyres on 6in-rim cast-alloy wheels, 205/70HR-14in rear tyres on 7in rims.
Dimensions: Wheelbase 8ft; front track 4ft 11.5in; rear track 4ft 11.5in. Length 13ft 9in; width 6ft 1.25in; height 3ft 7.7in. Unladen weight 2,218lb. Maximum payload 500lb.
Original UK price: (Mid-1976) £7,883 including taxes.
Differences for USA/Japanese markets:
Engine: CR 8.4:1. 2 Zenith-Stromberg CD single-choke carbs. 140bhp (DIN) at 6,500rpm. Maximum torque 130lb.ft at 5,000rpm.
Dimensions: Length 13ft 11.7in. Unladen weight 2,350lb.

Esprit S2 – produced 1978–80
Specification as for Esprit 1 except for:
Suspension and brakes: Front wheels had 7in rim width, rear wheels 7.5in rim width; special spare wheels had 185/70HR-13in tyre and 5.5in rim width.

Dimensions: Unladen weight 2,334lb.
Original price: (August 1978) £11,124 including taxes.

Esprit S2.2 – produced 1980–1
Specification as for Esprit S2 except for:
Engine: 95.28 x 76.2mm, 2,174cc, CR 9.4:1. 160bhp (DIN) at 6,500rpm. Maximum torque 160lb.ft at 5,000rpm.
Original UK price: (May 1980) £14,951 including taxes.

Esprit S3 – produced 1981–6
Specification as for Esprit S2.2 except for:
Suspension and brakes: Ifs as before; irs by coil springs, lower wishbones/semi-trailing arms, upper links and telescopic dampers. 10.5in front disc brakes, 10.8in rear disc brakes. Basic wheel/tyre equipment as before. Optional: 195/60VR-15in front tyres on 7in-rim alloy wheels, with 235/60VR-15in rear tyres on 8in-rim alloy wheels; special spare wheel with 175/70SR-14in tyre; 22.7mph/1,000rpm in top gear.
Dimensions: With optional wheels/tyres, front track 5ft 0.5in, rear track 5ft 1.2in. Unladen weight 2,352lb.
Original UK price: (April 1981) £13,461 including taxes.

Esprit HC – produced 1986–7
Specification as for Esprit S3 except for:
Engine: CR 10.9:1. 172bhp (DIN) at 6,500rpm. Maximum torque 163lb.ft at 5,000rpm.
Original UK price: (March 1987) £19,590 including taxes.

Esprit Turbo – produced 1980–6
Basic layout as for Esprit S3 except for:
Engine: 95.28 x 76.2mm, 2,174cc, CR 7.5:1, 2 twin-choke Dellorto DHLA carbs and Garrett AiResearch T3 turbocharger. 210bhp (DIN) at 6,000rpm. Maximum torque 200lb.ft at 4,000rpm.
Transmission: As S3 except for: 22.7mph/1,000rpm in top gear.

Suspension and brakes: As S3 but with 15in wheel/tyre equipment standard.

Dimensions: Unladen weight 2,653lb; maximum payload 466lb.

Original UK price: (February 1980 in 'Essex' form) £20,950 including taxes, (April 1981 production form) £16,982 including taxes.

For USA and related markets (HCPI) with fuel injection: 215bhp at 6,250rpm; maximum torque 192lb.ft at 5,000rpm.

Esprit Turbo HC – produced 1986–7

Specification as for Esprit Turbo of 1980–6 except for:

Engine: CR 8:1. 215bhp (DIN) at 6,000rpm. Maximum torque 220lb.ft at 4,250rpm.

Original UK price: (Autumn 1986) £24,980 including taxes.

Esprit – X180 style for 1988 model – produced 1987–90

Engine: 4-cyl, twin-overhead-camshaft, 95.28 x 76.2mm, 2,174cc, CR 10.9:1, 2 twin-choke Dellorto carbs. 172bhp (DIN) at 6,500rpm; maximum torque 163lb.ft at 5,000rpm.

Transmission: 5-speed all-synchromesh manual gearbox. Final drive ratio 3.88:1. Overall gear ratios 3.18, 4.00, 5.35, 7.95, 13.037, reverse 12.22:1. 23.7mph/1,000rpm in top gear.

Suspension and brakes: Ifs, coil springs, wishbones, anti-rollbar and telescopic dampers; irs, coil springs, lower wishbones/semi-trailing arms, upper links and telescopic dampers. Rack-and-pinion steering (with power-assistance). 10.1in front disc brakes, 10.8in rear disc brakes, with vacuum servo assistance. 195/60VR-15in front tyres on 7in-rim cast-alloy wheels, 235/60VR-15in rear tyres on 8in-rim cast-alloy wheels.

Dimensions: Wheelbase 8ft 0.76in; front track 5ft; rear track 5ft 1.2in. Length 14ft 2.5in; width 6ft 1.2in; height 3ft 8.8in. Unladen weight 2,590lb.

Original UK price: (Autumn 1989) £22,950 including taxes.

Esprit Turbo – new X180 style for 1988 model – produced 1987–90

Basic specification as for 1988 Esprit except for:

Engine: CR 8:1, 2 twin-choke Dellorto carbs with Garrett AiResearch turbocharger. 215bhp (DIN) at 6,500rpm. Maxi-mum torque 220lb.ft at 4,250rpm.

Dimensions: Unladen weight 2,800lb.

Original UK price: (Autumn 1987) £28,900 including taxes.

As Esprit Turbo (HCPI) for USA and related markets (1987–8), with Bosch fuel injection and old-type Citroen gearbox: 215bhp (DIN) at 6,250rpm; maximum torque 192lb.ft at 5,000rpm.

As Esprit Turbo (MPFI) for USA and related markets (1988–91), with multi-point Lucas/Delco fuel-injection: 228bhp (DIN) at 6,500rpm; maximum torque 218lb.ft at 4,000rpm.

Continued as Esprit ('Turbo' name omitted) in UK market from 1990.

Esprit Turbo SE – X180 style, produced from 1989

Basic specification as for original X180 Turbo except for:

Engine: Lucas/Delco fuel injection with turbocharger and intercooler; 264bhp (DIN) at 6,500rpm; maximum torque 261lb.ft at 3,900rpm.

Dimensions: 215/50ZR-15in front tyres on 7in rims, 245/50ZR-16in rear tyres on 8.5in rims.

Original UK price: (Mid-1989): £42,500 including taxes.

SE designation not used after 1990. Sold in USA in this specification from 1990.

Esprit S – produced 1990–1

Basic specification as for Esprit Turbo SE of 1989–90, for sale only in UK, except for: No intercooler. Maximum power 228bhp (DIN) at 6,500rpm; maximum torque 218lb.ft at 4,000rpm.

Esprit 2-litre – produced from 1991

For sale only in Italy, basic specification as for 1991 turbo-charged cars except for:

Engine: 95.28 x 69.24mm, 1,973cc, Lucas/Delco multi-point fuel injection; maximum power 250bhp (DIN) at 6,250rpm; maximum torque 226lb.ft at 3,750rpm.

Limited-edition models

JPS Commemorative – produced 1978–9

Based on the Esprit S2. To celebrate Team Lotus winning

Formula 1 World Championships for Drivers *and* Constructors. Each car was black with gold striping, plus special steering wheel and commemorative plaque.

Announced in October 1978 – 94 UK-spec and 53 export cars were built.

Essex Turbo Esprit – produced 1980-1
Based on the Esprit Turbo. Special build of 100 cars in Essex livery, the first cars produced with the turbocharged engine, to commemorate Essex Petroleum's sponsorship of the Team Lotus F1 cars.

Featuring dry-sump lubrication, air-conditioning and a Panasonic radio cassette player in the windscreen header panel.

58 cars built on the assembly line, but additional Esprit Turbos built and later converted to 'Essex' specification.

Limited Edition Esprit Turbo – produced 1986-7
Based on the Esprit Turbo HC of the period. To celebrate Lotus' 20th year at Hethel.

Available in three special two-tone colour combinations: Cherry Red/Calypso Red, Essex Blue/Mediterranean Blue, Shadow Grey/Silver Frost. All cars with full-leather interior and pigskin suede.

USA market Esprit Turbo – produced 1988
Based on the Esprit Turbo (X180 – HCPI). Built in Pearlescent White, with aerofoil-section rear wing, open-back tailgate, glass sunroof. A Sony 10-disc CD autochanger system was fitted in the USA by Lotus Cars USA.

88 cars produced.

40th Anniversary Limited Edition – produced 1988
Based on the Esprit Turbo (Carburettor *and* MPFI versions). Announced at the 1988 Birmingham motor show. To commemorate the 40th anniversary of the building of the first Lotus car.

Sold in all markets *except* the USA. All cars built in Pearlescent White except one car in Calypso Red. Fitted with a revised rear wing, glassback rear window and black bib spoiler.

40 cars produced, each with individually numbered silver dashboard plaque.

Indianapolis 25th Anniversary Esprit – produced 1990
Based on 1991-model Esprit SE. To commemorate Jim Clark's Lotus victory in the 1965 Indianapolis 500 race. All cars built in British Racing Green with yellow side stripes and detailing. Fitted with stainless steel dashboard plaques. Customers also given a framed replica of the 1965 Lotus 38 race car steering wheel.

25 cars built, all for sale in the USA market.

Esprit X180R – produced 1990
Based on the 1991-model Esprit SE, these cars built for the USA market to celebrate Lotus' successful first season of sportscar racing in the SCCA Escort World Challenge series.

The X180R was as close as possible to racing specification, while meeting all road car legality requirements. Extensive modifications included: blueprinted engine with revised inlet/exhaust systems; uprated transmission with limited-slip differential; Revolution three-piece wheels (8.5J x 16 fronts, 9.5J x 16 rears); Goodyear Eagle tyres (225/45ZR-16in fronts, 265/50ZR-16in rears); fully adjustable racing-type suspension; AP racing brakes (Lotus-Carlton/Omega type), with ABS.

Revised high-downforce aerodynamics; competition-style interior and instruments; factory-fitted rollcage and strengthened chassis.

20 cars built, all in Monaco White. A set of racing decals was provided loose, to be fitted if requested by the customer.

APPENDIX B
Production of Esprits from 1976

This section shows how production of Lotus Esprits fluctuated so much over the years. The Esprit remained in production at the end of 1992, though production was at a low rate.

For interest, a summary has been included of total Hethel production figures of Lotus cars since 1968, which was the first year in which it was complete.

* Eclat production began in the autumn of 1975.
** Esprit production began in the summer of 1976.
*** Elan (front-wheel drive) production began at the end of 1989.
**** Elan (front-wheel drive) production ended in July 1992.

Lotus annual production – 1968–92

Calendar year	Elan/Europa production	Elite/Eclat Excel/Esprit/ FWD Elan production	Total production
1968	3,048	–	3,048
1969	4,506	–	4,506
1970	3,373	–	3,373
1971	2,682	–	2,682
1972	2,996	–	2,996
1973	2,822	6	2,828
1974	759	687	1,446
1975	57	479*	535
1976	–	926**	926
1977	–	1,070	1,070
1978	–	1,200	1,200
1979	–	1,031	1,031
1980	–	383	383
1981	–	345	345
1982	–	541	541
1983	–	642	642
1984	–	837	837
1985	–	813	813
1986	–	703	703
1987	–	799	799
1988	–	1,302	1,302
1989	–	1,074***	1,074
1990	–	2,178	2,178
1991	–	2,241	2,241
1992	–	688****	688

The next table shows the individual production of each type of Esprit since 1976, in much more detail. As far as possible, the figures have been split into the five different model types; the trends revealed are fascinating:

Calendar year	Esprit (normally-aspirated)	Esprit Turbo	Esprit Turbo USA*	Esprit Turbo USA**	Esprit Turbo Inter-cooled
1976	134	–	–	–	–
1977	580	–	–	–	–
1978	553	–	–	–	–
1979	474	–	–	–	–
1980	80	57	–	–	–
1981	185	116	–	–	–
1982	160	205	–	–	–
1983	84	343	–	–	–
1984	104	418	–	–	–
1985	127	262	62	–	–
1986	72	136	246	–	–
1987	78	186	198	–	–
1988	176	387	375	120	–
1989	90	103	–	121	563
1990	22	51	–	8	698
1991	–	9	–	3	113
1992	–	1	–	–	172
Totals	2,919	2,274	881	252	1,546

Notes: Although the Esprit was first shown to the public in October 1975, the first deliveries were not made until June 1976.
The first 100 Esprit Turbos (1980 and 1981) were decorated as 'Essex Commemorative' models, with distinctive colour schemes.
Assembly of normally-aspirated Esprits ended in 1990.

Assembly of intercooled Esprit Turbos took over completely in 1991.

*　This model was specially engineered for sale in the USA.
**　This model was the restyled-shape car, specially engineered for sale in the USA.

APPENDIX C

Performance figures for the mid-engined Lotus

As with the front-engined Lotus cars covered in the companion volume, there was a great deal of choice when compiling this Appendix.

Because I believe them to provide the most accurate coverage of all cars, I am grateful to my good friends at *Autocar* – now *Autocar & Motor* – for permission to quote performance figures.

The single USA-specification test was published by the similarly worthy *Road & Track*.

Much controversy still surrounds the true performance of the S1 Esprit, as *Autocar*'s test car certainly produced a disappointing top speed and was not as accelerative as Lotus claimed it should have been. (*Motor*'s test car – a different example – also had a disappointing top speed, and its standing-start figures are in any case suspect as the quarter-mile time is almost the same as that recorded by *Autocar* while the 0–80mph figure is 2.5sec quicker!).

Accordingly, the S2 should be regarded as the representative example of 2-litre Esprit performance. No-one, by the way, is complaining about the pace of the S3 or the Turbo derivatives!

Notes concerning performance table on next page:

*　The test car could not exceed 124mph, when Lotus claimed that 138mph was achieved, and all the evidence points to a lack of top-end power, for acceleration at lower rpm is more closely related to that claimed by Lotus.

**　This figure is quoted in US miles per gallon – and gives an Imperial equivalent of around 33, which is mind-bogglingly economical; the US open-road speed limits have a lot to answer for! *Road & Track*, who published this test, do not observe top-gear acceleration figures.

***　I am deeply suspicious of the quoted weight for this *Autocar & Motor* test car, which at 3,052lb must surely be a mistake. It should have been about 2,650lb.

	Esprit S1 1,973cc 160bhp (DIN)	Esprit S2 1,973cc 160bhp (DIN)	Esprit S3 2,174cc 160bhp (DIN)	Esprit Turbo 2,174cc 210bhp (DIN)	Esprit S1 (USA) 1,973cc 140bhp (DIN)	Esprit Turbo HC 2,174cc 215bhp (DIN)	Esprit Turbo (new style) 2,174cc 215bhp (DIN)
Mean maximum speed (mph)	124*	135	134	148	120	141	150
Acceleration (sec)							
0–30mph	2.9	2.8	2.3	2.3	3.1	2.0	2.3
0–40mph	4.3	4.2	3.5	3.2	4.5	2.9	3.3
0–50mph	6.3	5.9	5.0	4.7	6.7	4.4	4.3
0–60mph	8.4	8.0	6.7	6.1	9.2	5.6	5.4
0–70mph	11.6	10.7	9.6	8.3	12.3	7.7	7.1
0–80mph	15.3	13.8	12.4	10.3	16.2	9.5	8.7
0–90mph	20.2	17.4	16.1	13.0	21.7	12.0	10.7
0–100mph	27.4	22.7	20.9	17.0	–	15.0	13.3
0–110mph	39.4	30.7	27.9	20.7	–	18.5	16.3
0–120mph	–	–	40.7	27.1	–	23.1	20.1
0–130mph	–	–	–	39.2	–	–	26.0
Standing ¼-mile (sec)	16.3	16.0	15.5	14.6	17.0	14.4	13.7
Top gear accelereration (sec)							
20–40mph	–	15.4	–	14.9	–	12.0	12.6
30–50mph	13.9	14.4	10.4	10.4	–	9.8	9.6
40–60mph	12.4	13.6	10.0	8.3	–	8.0	7.2
50–70mph	12.9	12.9	9.9	8.5	–	6.8	6.3
60–80mph	12.5	13.8	10.0	8.4	–	6.5	6.3
70–90mph	13.4	15.7	11.2	8.7	–	6.6	6.3
80–100mph	17.0	16.6	12.9	9.2	–	7.1	6.3
90–110mph	23.6	21.1	16.9	10.5	–	8.0	6.9
100–120mph	–	–	–	13.1	–	9.3	7.7
110–130mph	–	–	–	19.3	–	–	9.3
Overall fuel consumption (mpg Imp)	23.3	19.4	21.7	18.0	27.5**	20.9	19.6
Typical fuel consumption (mpg Imp)	26	21	24	20	–	23	–
Kerb weight (lb)	2,275	2,334	2,489	2,653	2,480	2,530	Not clear***
Original test published	Jan 1977	Jan 1979	Jun 1981	May 1981	July 1977	Apr 1987	Apr 1988